Cambridge Tracts in Mathematics
and Mathematical Physics

GENERAL EDITORS
P. HALL, F.R.S., AND F. SMITHIES, PH.D.

No. 45

AN INTRODUCTION TO
DIOPHANTINE APPROXIMATION

T0275840

AN INTRODUCTION TO
DIOPHANTINE APPROXIMATION

BY

J. W. S. CASSELS

PH.D.

*Fellow of Trinity College, Cambridge
and Lecturer in Mathematics in the
University of Cambridge*

CAMBRIDGE
AT THE UNIVERSITY PRESS
1965

CAMBRIDGE UNIVERSITY PRESS
Cambridge, New York, Melbourne, Madrid, Cape Town, Singapore, São Paulo, Delhi

Cambridge University Press
The Edinburgh Building, Cambridge CB2 8RU, UK

Published in the United States of America by Cambridge University Press, New York

www.cambridge.org
Information on this title: www.cambridge.org/9780521045872

First published 1957
Reprinted 1965
Re-issued in this digitally printed version 2008

A catalogue record for this publication is available from the British Library

ISBN 978-0-521-04587-2 paperback

CONTENTS

PREFACE

This tract sets out to give some idea of the basic techniques and of some of the most striking results of Diophantine approximation to, say, an undergraduate in his final year with no knowledge of number theory beyond the rudiments. It is self-contained for such a reader except that the elements of Lebesgue theory are required in Chapter VII and the elements of the theory of algebraic numbers are required in Chapter VIII (but not in Chapter VI). What is needed from the geometry of numbers is developed in detail in Appendix B, which the reader should consult as becomes necessary.

Bibliographical notes and suggestions for further reading are kept to the end of the chapters and occasionally comments there are for a more sophisticated reader than is the text. In general, I have mentioned only the most recent or most accessible papers from which further references may be obtained. For work up to 1936 there is the masterly and indispensable report of KOKSMA (1936).† Where no reference is given it should not be assumed that originality is claimed—there is much that is common property or that has been acquired without my remembering the source.

The expert will recognize gaps. In particular, there is very little about simultaneous approximation to a number of irrationals and nothing about the exact constants. The little precise knowledge that we have here depends on deep results, e.g. Davenport's value of the critical determinant of $|X|(Y^2+Z^2) \leqslant 1$, and quite different techniques from those discussed here (indeed I do not use the word 'lattice' at all). For the present state of knowledge see DAVENPORT (1954). However, the reader will find in Appendix B the prerequisites for the understanding of Mahler's compactness theorem for lattices (MAHLER (1946)) which is an essential tool for discussing simultaneous approximation as well as in many other contexts.

There are analogues of many of the results of this tract in which p-adic numbers take the role of the real numbers; see LUTZ (1951) and the literature cited there.

† For references see p. 161.

I should like here to express my gratitude to those who have helped me. Professors H. Davenport, K. Mahler, L. J. Mordell and Mr B. J. Birch have read both the original manuscript and the proofs; and Professor P. Hall and Mr H. P. F. Swinnerton-Dyer have read the proofs: that the final tract resembles so little the original conception is largely because of their penetrating criticism both of form and content. Professor C. A. Rogers and Mr B. J. Birch have allowed me to use unpublished work relating to the Markoff Chain and to transference theorems respectively, and Dr K. F. Roth put a manuscript of his revolutionary improvement of the Thue-Siegel Theorem at my disposal before publication.

J. W. S. C.

Trinity College
CAMBRIDGE
1956

NOTATION

1. 'Number' means 'real number' unless the contrary is explicitly stated or necessarily implied by the context.

2. For a number θ the following usages are standard:

$[\theta]$ is the integral part of θ; that is, the integer such that $[\theta] \leqslant \theta < [\theta] + 1$.

$\{\theta\}$ is the fractional part of θ; that is, $[\theta] + \{\theta\} = \theta$.

$\|\theta\|$ is the difference, taken positively, between θ and the nearest integer; that is,

$$\|\theta\| = \min(\{\theta\}, 1 - \{\theta\}) = \min |\theta - n| \quad (n = 0, \pm 1, \pm 2, \ldots).$$

Clearly $\qquad \|\theta_1 + \theta_2\| \leqslant \|\theta_1\| + \|\theta_2\|,$

and $\qquad\qquad \|n\theta\| \leqslant |n| \|\theta\|,$

for integers n.

Brackets [ ], { } are not otherwise used except where there can be no fear of ambiguity.

3. Vectors (ordered sets of numbers) are denoted by bold-faced type and their elements by the corresponding ordinary letters, e.g. $\boldsymbol{\alpha} = (\alpha_1, \ldots, \alpha_n)$, $\mathbf{z} = (z_1, \ldots, z_m)$. If a sequence of vectors (of the same number of elements) has to be distinguished by affixes it is done as follows:

$$\mathbf{z}^{(r)} = (z_{r1}, \ldots, z_{rm}) \quad (r = 1, 2, \ldots).$$

The zero vector $(0, \ldots, 0)$ of any number of elements is denoted by $\mathbf{0}$. The usual notation for the addition and multiplication of vectors is used:

$$\lambda \mathbf{z} = (\lambda z_1, \ldots, \lambda z_m),$$
$$\mathbf{z}^{(1)} + \mathbf{z}^{(2)} = (z_{11} + z_{21}, \ldots, z_{1m} + z_{2m}).$$

If $\mathbf{u} = (u_1, \ldots, u_m)$ and $\mathbf{z} = (z_1, \ldots, z_m)$ we put

$$\mathbf{u}\mathbf{z} = u_1 z_1 + \ldots + u_m z_m.$$

We do not try to make a distinction between covariant and contravariant vectors, even where it might be appropriate to do so.

We often think of vectors as points in the appropriate Euclidean space and use the natural language to express their relations to one another.

4. $a|b$ for integers a, b means 'a divides b'. Similarly $a \nmid b$ means 'a does not divide b'. The greatest common divisor and the least common multiple of a set of integers $a_1, ..., a_m$ are written g.c.d. $(a_1, ..., a_m)$ or g.c.d. (a_j) and l.c.m. $(a_1, ..., a_m)$ or l.c.m. (a_j) respectively.

5. We borrow the symbol ϵ from the logicians. If A is a set of things and a is a thing then $a \epsilon A$ means 'a belongs to A'. For example if A is the set of rational numbers and a is a number then $a \epsilon A$ means 'a is a rational number'. $\notin$ is the opposite of ϵ.

6. The least upper bound and the greatest lower bound of a set A of real numbers a are denoted by $\sup_{a \epsilon A} a$, $\inf_{a \epsilon A} a$ respectively. The upper and lower limit of a sequence a_j of real numbers are denoted by $\limsup a_j$ and $\liminf a_j$ respectively; so

$$\liminf a_j = \lim_{J \to \infty} (\inf_{j \geqslant J} a_j).$$

7. Theorems and lemmas are numbered consecutively in each chapter but equations etc. only in each section. A reference (7) means 'expression (7) of the present section' while (2·7) means 'expression (7) of §2'.

8. A list of works referred to is given on p. 161. These are quoted by giving the author's name in small capitals and the year, e.g. PERRON (1913). Works in the same year are distinguished by small letters a, b.

9. Some words or phrases defined in the text are listed on p. 166. They are set in small capitals where they are defined.

CHAPTER I

HOMOGENEOUS APPROXIMATION

1. Introduction. In this chapter and the next we shall consider how closely (in a suitable sense) an irrational number θ may be approximated to by rational fractions p/q. Here p, q are integers and, without loss of generality, it may be supposed that $q > 0$. Since for fixed q the minimum of $|\theta - p/q| = q^{-1}|q\theta - p|$ is† $q^{-1}\|q\theta\|$, we may consider $\|q\theta\|$ instead of $|\theta - p/q|$. This leads naturally to the discussion of continued fractions, which are useful tools.

A simple but useful result is

THEOREM I. *Let θ and $Q > 1$ be real. Then there is an integer q such that*
$$0 < q < Q, \quad \|q\theta\| \leqslant Q^{-1}.$$

Note 1. Here θ is not necessarily irrational. The theorem thus gives information on the approximation of rational numbers by rational numbers with smaller denominators.

Note 2. If Q is an integer and $\theta = Q^{-1}$ then $\|q\theta\| \geqslant Q^{-1}$ for $0 < q < Q$. Hence the $\leqslant$ in the theorem cannot be improved to $<$.

First proof. (Dirichlet.) Suppose first that Q is an integer. Consider the distribution of the $Q + 1$ numbers†

$$0, \quad 1, \quad \{q\theta\} \qquad (0 < q < Q), \tag{1}$$

which all satisfy $0 \leqslant x \leqslant 1$, into the Q subintervals

$$\frac{u}{Q} \leqslant x < \frac{u+1}{Q}, \quad 0 \leqslant u < Q, \tag{2}$$

where $\leqslant$ is to be read for $<$ when $u = Q - 1$. At least one of the subintervals (2) must contain two of the points (1); that is, we may find integers r_1, r_2, s_1, s_2 such that

$$|(r_1\theta - s_1) - (r_2\theta - s_2)| \leqslant Q^{-1},$$

where $r_2 < r_1$ without loss of generality. Then $q = r_1 - r_2$ does what is required.

† For notation see p. ix.

The truth of the theorem when Q is not an integer follows at once from its truth for $[Q]+1$.

Second proof. The theorem is essentially a special case of Minkowski's linear forms theorem (Theorem III of Appendix B). By it there are integers p, q not both 0 such that

$$|\theta q - p| \leqslant Q^{-1}, \quad |q| < Q.$$

If $q = 0$ we should have $|p| = |\theta q - p| \leqslant Q^{-1} < 1$ and $p = 0$. Hence, by taking $-p$, $-q$ for p, q if need be, we may suppose that $q > 0$.

2. The continued fraction process.

By Theorem I there are infinitely many integer solutions $q > 0$ of

$$q \| q\theta \| < 1, \tag{1}$$

if θ is irrational. The continued fraction process gives more detailed information, and, in particular, will allow us to improve the 1 in the last statement to $5^{-\frac{1}{2}}$. The process is a fundamental one and plays a basic part in many investigations, though we shall not use it much in this book.

A fraction p/q $(q > 0)$ gives a BEST APPROXIMATION† to θ if

$$\| q\theta \| = | q\theta - p |,$$

and if $\qquad \| q'\theta \| > \| q\theta \| \quad$ for $\quad 0 < q' < q.$

Clearly $q = q_1 = 1$ gives a best approximation with some $p = p_1$ and

$$| q_1\theta - p_1 | = \| \theta \| \leqslant \tfrac{1}{2}.$$

If $\| q_1\theta \| = 0$, i.e. if θ is an integer, the process stops. If not, there is certainly a value of q such that $\| q\theta \| < \| q_1\theta \|$ (e.g. by Theorem I with any $Q > \| q_1\theta \|^{-1}$). Let q_2 be the least q with this property so that $| q_2\theta - p_2 | = \| q_2\theta \| < \| q_1\theta \|$ for some p_2, but $\| q\theta \| \geqslant \| q_1\theta \|$ for $0 < q < q_2$. If $\| q_2\theta \| = 0$, the process stops. If not, we can continue the process and find a sequence of integers‡

$$q_1 = 1 < q_2 < q_3 < \ldots,$$

and $p_1, p_2, \ldots$, such that

$$\| q_n\theta \| = | q_n\theta - p_n |, \tag{2}$$

$$\| q_{n+1}\theta \| < \| q_n\theta \|, \tag{3}$$

$$\| q\theta \| \geqslant \| q_n\theta \| \quad \text{for} \quad 0 < q < q_{n+1}. \tag{4}$$

---

† For this use of small capitals see pp. x, 166.
‡ We shall change this notation slightly later (p. 5).

From (4) and Theorem I with $Q = q_{n+1}$ we have

$$q_n \| q_n \theta \| < q_{n+1} \| q_n \theta \| \leqslant 1. \tag{5}$$

If $q_{n+1} \theta - p_{n+1}$ and $q_n \theta - p_n$ both had the same sign, we should have

$$| q' \theta - p' | < | q_n \theta - p_n |,$$

where $p' = p_{n+1} - p_n$, $0 < q' = q_{n+1} - q_n < q_{n+1}$, in contradiction to (4). Hence

$$(q_n \theta - p_n)(q_{n+1} \theta - p_{n+1}) \leqslant 0. \tag{6}$$

LEMMA 1 A. *The p_n/q_n are just all the best approximations to θ in order of ascending q_n.*
B. *If θ is rational then $\theta = p_N/q_N$ for some N.*
C. *If θ is irrational then $p_n/q_n \to \theta$.*
Proof A. By construction p_{n+1}/q_{n+1} is the best approximation p/q with smallest $q > q_n$.
B. If $\theta = u/v$ $(v > 0)$ in its lowest terms then u/v is clearly a best approximation.
C. $| \theta - p_n/q_n | < q_n^{-2}$ by (5).

LEMMA 2. $q_{n+1} p_n - q_n p_{n+1} = \pm 1$.
Proof. The left-hand side is an integer, and

$$q_{n+1} p_n - q_n p_{n+1} = q_n (q_{n+1} \theta - p_{n+1}) - q_{n+1} (q_n \theta - p_n). \tag{7}$$

Hence by (5), (6), we have

$$| q_{n+1} p_n - q_n p_{n+1} | = q_n \| q_{n+1} \theta \| + q_{n+1} \| q_n \theta \| > 0$$

and

$$< 2 q_{n+1} \| q_n \theta \| \leqslant 2. \tag{7'}$$

COROLLARY 1. $q_{n+1} p_n - q_n p_{n+1}$ *has the opposite sign to* $q_n \theta - p_n$.
COROLLARY 2. $q_{n+1} p_n - q_n p_{n+1} = -(q_n p_{n-1} - q_{n-1} p_n)$.
COROLLARY 3. $q_n \| q_{n+1} \theta \| + q_{n+1} \| q_n \theta \| = 1$.
Proofs. Clear, by (6), (7) and (7').

LEMMA 3. *For $n \geqslant 2$ there is an integer $a_n \geqslant 1$ such that*

$$q_{n+1} = a_n q_n + q_{n-1}, \tag{8}$$

$$p_{n+1} = a_n p_n + p_{n-1}, \tag{9}$$

$$| q_{n-1} \theta - p_{n-1} | = a_n | q_n \theta - p_n | + | q_{n+1} \theta - p_{n+1} |. \tag{10}$$

Proof. By Lemma 2, Corollary 2, we have

$$p_n(q_{n+1} - q_{n-1}) = q_n(p_{n+1} - p_{n-1}).$$

Hence $q_{n+1} - q_{n-1} = a_n q_n$, $p_{n+1} - p_{n-1} = a_n p_n$ for some integer a_n, since p_n, q_n are coprime by Lemma 2 (or by the definition of a best approximation). Here $a_n > 0$ since $q_{n+1} > q_{n-1}$. Finally (10) follows from (8) and (9), by (6).

(10) gives a simple procedure for finding p_{n+1}, q_{n+1} once the p_ν, q_ν ($\nu \leqslant n$) are known. If θ is rational and $\| q_n \theta \| = 0$, the process stops with p_n, q_n. Otherwise†

$$a_n = \left[\frac{\| q_{n-1} \theta \|}{\| q_n \theta \|} \right],$$

by (2), (10) since $\| q_{n+1} \theta \| < \| q_n \theta \|$; and then p_{n+1}, q_{n+1} can be found from (8) and (9).

To start this process we must know q_2 as well as $q_1 = 1$. We can assume for simplicity from now on that

$$0 < \theta < 1,$$

since the addition of an integer to θ does not affect the q_n and affects the p_n trivially. Suppose first that

$$0 < \theta \leqslant \tfrac{1}{2}.$$

Then $q_1 \theta - p_1 = \theta > 0$, $q_1 = 1$, $p_1 = 0$.

Hence by Lemma 2 and its first corollary we have $p_2 = 1$. Thus Lemma 3 continues to hold for $n = 1$ if we put

$$p_0 = 1, \quad q_0 = 0, \quad a_1 = q_2.$$

In particular (10) for $n = 1$ becomes

$$1 = a_1 \theta + \| q_2 \theta \|$$

and so $a_1 = [\theta^{-1}]$. Suppose, secondly, that

$$\tfrac{1}{2} < \theta < 1.$$

Then $q_1 \theta - p_1 = \theta - 1 < 0$, $q_1 = p_1 = 1$, and so $q_2 - p_2 = 1$ by Lemma 2 and its first corollary. To preserve the scheme of Lemma 3 with the same start as before we must take

$$p_{-1} = 1, \quad q_{-1} = 0,$$
$$p_0 = 0, \quad q_0 = 1, \quad a_0 = 1,$$
$$p_1 = 1, \quad q_1 = 1, \quad a_1 = q_2 - 1 = p_2.$$

Then (10) becomes for $n = 0, 1$

$$1 = \theta + |\theta - 1| \quad (n = 0),$$
$$\theta = a_1 |\theta - 1| + |q_2\theta - p_2| \quad (n = 1),$$

where

$$1 > \theta = |q_0\theta - p_0| > |\theta - 1| = |q_1\theta - p_1| > |q_2\theta - p_2|.$$

It is convenient to change the notation if $\frac{1}{2} < \theta < 1$ to have a consistent start:†

THEOREM II. *Let $0 < \theta < 1$ and let integers p_n, q_n, a_n be defined by*

(A)
$$\left. \begin{array}{ll} p_0 = 1, & q_0 = 0 \\ p_1 = 0, & q_1 = 1 \end{array} \right\},$$

(B)
$$\left. \begin{array}{l} p_{n+1} = a_n p_n + p_{n-1} \\ q_{n+1} = a_n q_n + q_{n-1} \end{array} \right\} \quad (n \geqslant 1),$$

where
$$a_n = \left[\frac{|q_{n-1}\theta - p_{n-1}|}{|q_n\theta - p_n|} \right]$$

if $q_n\theta \neq p_n$: and the process stops with p_n, q_n if $q_n\theta = p_n$. Then the p_n/q_n are the best approximations to θ for $n \geqslant 1$ if $a_1 > 1$ and for $n \geqslant 2$ if $a_1 = 1$. Further,

$$(-)^{n+1}(q_n\theta - p_n) \geqslant 0$$

and
$$q_{n+1}p_n - q_n p_{n+1} = (-1)^n.$$

Proof. This is an immediate consequence of what precedes. The signs in the last two expressions follow from their values when $n = 1$ and from (6) or Lemma 2, Corollary 2. It is usual to speak of the p_n/q_n as the CONVERGENTS to θ (whether or not they give best approximations) and to call the a_n the PARTIAL QUOTIENTS.

Since the a_n are determined by θ and since, by Lemma 1, θ is determined by the a_n, we may write without fear of ambiguity

$$\theta = [a_1, a_2, a_3, \ldots],$$

† Many writers use a slightly different notation, see the notes at the end of the chapter.

if θ is irrational, and
$$\theta = [a_1, \ldots, a_N],$$
if $\theta = p_{N+1}/q_{N+1}$.

Write
$$\theta_0 = 1, \quad \theta_n = \frac{|q_n\theta - p_n|}{|q_{n-1}\theta - p_{n-1}|} \quad (n \geqslant 1),$$
so that
$$\theta_1 = \theta, \quad 0 \leqslant \theta_n < 1 \quad (n \geqslant 1). \tag{11}$$

Then (10) becomes
$$\theta_n^{-1} = a_n + \theta_{n+1}. \tag{12}$$

In particular, $\theta_{N+1} = 0$ if $\theta = p_{N+1}/q_{N+1}$. A rational θ thus has the form

$$\theta = \cfrac{1}{a_1 + \cfrac{1}{a_2 + \cfrac{1}{\ddots \cfrac{}{a_{N-1} + \cfrac{1}{a_N}}}}}, \tag{13}$$

which explains the name continued fraction. Now $a_N^{-1} = \theta_N < 1$ by (11), (12), i.e. $a_N \neq 1$. It is useful, however, to define

$$[a_1, \ldots, a_{N-1}, 1] = [a_1, \ldots, a_{N-1} + 1]; \tag{14}$$

but the second expression determines the p_n/q_n which are best approximations (if the first is used, p_N/q_N is not a best approximation).

Since the a_n are determined by (12) we see that

$$\theta_n = [a_n, a_{n+1}, \ldots] \text{ or } [a_n, \ldots, a_N].$$

Similarly write
$$\phi_n = q_n/q_{n+1} \quad (n \geqslant 0),$$
so that
$$0 \leqslant \phi_n \leqslant 1,$$
with equality only in easily identifiable cases. Then

$$q_{n+1} = a_n q_n + q_{n-1}$$
becomes
$$\phi_n^{-1} = a_n + \phi_{n-1}.$$

Hence, as with θ_n, we have

$$\phi_n = [a_n, a_{n-1}, ..., a_1].$$

We may now express $q_n \| q_n \theta \|$ in terms of the θ_n and ϕ_n. By Lemma 2, Corollary 3,

$$1 = q_{n+1} \| q_n \theta \| + q_n \| q_{n+1} \theta \|$$
$$= (\phi_n^{-1} + \theta_{n+1}) q_n \| q_n \theta \|$$
$$= (\phi_{n-1} + a_n + \theta_{n+1}) q_n \| q_n \theta \|.$$

Hence
$$q_n \| q_n \theta \| = (a_n + \theta_{n+1} + \phi_{n-1})^{-1} \qquad (15)$$

and
$$q_{n+1} \| q_n \theta \| = (1 + \theta_{n+1} \phi_n)^{-1} > \tfrac{1}{2}. \qquad (16)$$

We have already seen that every θ, $0 < \theta < 1$, determines a sequence $a_1, a_2, ...$, of positive integers. We now prove the converse that every sequence of positive integers determines a θ; but we need a trivial lemma.

LEMMA 4. *Let* $n \geqslant 1$ *and let*

$$\theta = [a_1, ..., a_n, a_{n+1}, a_{n+2}, ...],$$
$$\theta' = [a_1, ..., a_n, b_{n+1}, b_{n+2}, ...],$$

where the expressions may terminate. Then

$$| \theta - \theta' | < 2^{-(n-2)}. \qquad (17)$$

Note. All we really need is that the right-hand side of (17) tends to 0 as $n \to \infty$.

Proof. Let p_ν, q_ν $(0 \leqslant \nu \leqslant n+1)$ be defined by (A), (B) of Theorem II. Then by Theorem II p_{n+1}/q_{n+1} is a best approximation both to θ and θ'; and $q_{n+1}\theta - p_{n+1}$, $q_{n+1}\theta' - p_{n+1}$ have the same sign. But $| q_{n+1}\theta - p_{n+1} | < q_{n+1}^{-1}$, $| q_{n+1}\theta' - p_{n+1} | < q_{n+1}^{-1}$ by (5) and so $| \theta - \theta' | < q_{n+1}^{-2}$. But by induction $q_{n+1} > 2^{\frac{1}{2}(n-2)}$, since $q_{n+1} = a_n q_n + q_{n-1} > 2q_{n-1}$.

THEOREM III. *Let* $a_1, a_2, ..., a_N$ *or* $a_1, a_2, ...$ *be a finite or infinite sequence of positive integers. Then there is a* θ *such that* $\theta = [a_1, ..., a_N]$ *or* $\theta = [a_1, a_2, ...]$ *respectively. If* $a_N = 1$ *the interpretation* (14) *is to be taken.*

Proof. In the finite case put $\theta_{N+1} = 0$ and define

$$\theta_N, \theta_{N-1}, ..., \theta_1 = \theta$$

in order by (12). Clearly $0 < \theta_n \leqslant 1$ for $1 \leqslant n \leqslant N$ and $\theta_n = 1$ only if $n = N$, $a_N = 1$. Hence $\theta = [a_1, a_2, ..., a_N]$.

In the infinite case write

$$\theta^{(N)} = [a_1, ..., a_N],$$

which we now know to exist. By Lemma 4,

$$\lim |\theta^{(N)} - \theta^{(M)}| = 0 \quad (N \to \infty, M \to \infty),$$

and so $\theta = \lim \theta^{(N)} \geqslant 0$

exists. Similarly

$$\theta_n = \lim \theta_n^{(N)} \geqslant 0, \quad \theta_n^{(N)} = [a_n, ..., a_N],$$

exists. But now $(\theta_n^{(N)})^{-1} = a_n + \theta_{n+1}^{(N)},$

if $N > n+1$. Hence, in the limit $\theta_n^{-1} = a_n + \theta_{n+1}$ for all n, as required.

3. Equivalence. Two real numbers θ, θ' are EQUIVALENT if there are integers r, s, t, u, such that

$$\theta = \frac{r\theta' + s}{t\theta' + u}, \quad ru - ts = \pm 1.$$

Since $\theta' = \frac{-u\theta + s}{t\theta - r},$

the equivalence relation is a symmetrical one. Further, if θ is equivalent to θ' and θ' to θ'' then it is easily verified by direct substitution that θ is equivalent to θ''.

In the continued fraction process we have $\theta_n = (a_n + \theta_{n+1})^{-1}$. Hence each of $\theta = \theta_1, \theta_2, ...$ is equivalent to every other. More generally if

$$\theta = [a_1, ..., a_l, c_1, c_2, ...],$$
$$\theta' = [b_1, ..., b_m, c_1, c_2, ...],$$

then each is equivalent to

$$\theta_{l+1} = \theta'_{m+1} = [c_1, c_2, ...],$$

and so they are equivalent to each other. In particular, any two rational numbers are equivalent. We shall now prove

THEOREM IV. *The necessary and sufficient condition that* θ, θ' $(0 < \theta, \theta' < 1)$ *be equivalent is that*

$$\theta = [a_1, a_2, \ldots, a_l, c_1, c_2, \ldots],$$
$$\theta' = [b_1, b_2, \ldots, b_m, c_1, c_2, \ldots],$$

for suitable l, m *and* $a_1, \ldots, a_l, b_1, \ldots, b_m, c_1, c_2, \ldots.$

Proof. It remains only to show that if θ, θ' are irrational and equivalent they can be written in the form specified. Let

$$\theta = \frac{r\theta' + s}{t\theta' + u}, \quad ru - st = \pm 1. \tag{1}$$

Then identically
$$q\theta - p = \frac{q'\theta' - p'}{t\theta' + u}, \tag{2}$$

where
$$q' = qr - pt, \quad p' = -qs + pu. \tag{3}$$

Since $ru - st = \pm 1$, we may solve (3) to give

$$\pm q = q'u + p't, \quad \pm p = q's + p'r. \tag{4}$$

In what follows dashed and undashed pairs of symbols will always be related as (p, q), (p', q') are in (3), (4).

The first part of (3) may be written

$$q' = q(r - t\theta) + t(q\theta - p). \tag{5}$$

By changing the signs of r, s, t, u simultaneously, if necessary, we may suppose that
$$r - t\theta > 0.$$

Then (5) shows that the integer q' has the sign of the integer q provided that

$$|q\theta - p| < \frac{r - t\theta}{|t|}, \tag{6}$$

the right-hand side being independent of p and q.

Now let p_n/q_n, p_{n+1}/q_{n+1} be two successive best approximations to θ and let p'_n, q'_n, p'_{n+1}, q'_{n+1} be defined by (3), (4) (as explained above). We shall show that p'_n/q'_n, p'_{n+1}/q'_{n+1} are in fact two successive best approximations to θ' provided that n is large enough.

In the first place, both $(p, q) = (p_n, q_n)$ and $(p, q) = (p_{n+1}, q_{n+1})$ satisfy (6) if n is large enough and then $q_n' > 0$, $q_{n+1}' > 0$. Similarly $q_{n+1}' - q_n' > 0$ if n is large enough since then

$$(p, q) = (p_{n+1} - p_n, q_{n+1} - q_n)$$

satisfies (6) and $q_{n+1} - q_n > 0$. Hence

$$0 < q_n' < q_{n+1}'. \tag{7}$$

By (2),
$$\begin{aligned}
|q_n'\theta' - p_n'| &= |t\theta' + u|\,|q_n\theta - p_n| \\
&> |t\theta' + u|\,|q_{n+1}\theta - p_{n+1}| \\
&= |q_{n+1}'\theta' - p_{n+1}'|.
\end{aligned} \tag{8}$$

Now suppose that there exists a pair of integers (x', y') such that
$$0 < y' < q_{n+1}', \quad |y'\theta' - x'| \leqslant |q_n'\theta' - p_n'|. \tag{9}$$

Let (x, y) correspond to (x', y') by (3), (4). As in (8) we deduce from (9) that
$$|y\theta - x| \leqslant |q_n\theta - p_n|. \tag{10}$$

Hence both $(p, q) = (x, y)$ and $(p, q) = (p_{n+1} - x, q_{n+1} - y)$ satisfy (6) if n is large enough and then

$$0 < y < q_{n+1}, \tag{11}$$

by (9) (cf. proof of (7)). But (10), (11) imply $(x, y) = (p_n, q_n)$ since p_n/q_n, p_{n+1}/q_{n+1} are successive best approximations. Hence $(x', y') = (p_n', q_n')$ is the only integer solution of (9). This with (7), (8) shows that p_n'/q_n', p_{n+1}'/q_{n+1}' are successive best approximations to θ'.

For all $n \geqslant$ some N the p_n'/q_n' in order are thus the successive best approximations to θ': but p_n'/q_n' is not necessarily the nth. If $\theta = [a_1, a_2, \ldots]$, we have

$$q_{n+1}' = rq_{n+1} - tp_{n+1} = a_n q_n' + q_{n-1}'.$$

Hence $\theta' = [b_1, \ldots, b_s, a_{N+1}, a_{N+2}, \ldots]$ for some s and $b_1, \ldots, b_s$, as required.

For irrational θ put†

$$\nu(\theta) = \liminf q \| q\theta \|,$$

so that $0 \leqslant \nu(\theta) \leqslant 1$ by Theorem I. The inequality $q \| q\theta \| < \nu'$ has

---

† For the notation 'lim inf' see p. x.

infinitely many integer solutions $q > 0$ if $\nu' > \nu(\theta)$ but only a finite number of solutions if $\nu' < \nu(\theta)$. By (2·4) clearly

$$\nu(\theta) = \liminf q_n \| q_n \theta \|.$$

COROLLARY. $\nu(\theta) = \nu(\theta')$, if θ is equivalent to θ'.

Proof. Suppose that there are infinitely many solutions of

$$q \, | \, q\theta - p \, | < \kappa, \tag{12}$$

for some κ and let p', q' be defined by (3), (4). On interchanging the roles of θ, θ' in (2) we have

$$q'\theta' - p' = \pm \frac{q\theta - p}{r - t\theta},$$

the $\pm$ sign being that of (1) and (4). By this and (5),

$$q' \, | \, q'\theta' - p' \, | \leqslant q \, | \, q\theta - p \, | + \frac{|t|}{|r - t\theta|} \, | \, q\theta - p \, |^2$$

$$\leqslant \kappa + \frac{|t|}{|r - t\theta|} \cdot \left(\frac{\kappa}{q} \right)^2 < \kappa',$$

for any fixed $\kappa' > \kappa$ provided that q is large enough. Hence

$$q' \, | \, q'\theta' - p' \, | < \kappa'$$

has infinitely many solutions for any $\kappa' > \kappa$; that is $\nu(\theta') \leqslant \nu(\theta)$. Similarly $\nu(\theta) \leqslant \nu(\theta')$.

The corollary also follows readily from (2·15) and Lemma 4.

4. Application to approximations.

The continued fraction machinery rapidly proves

THEOREM V. Let θ be irrational. Then there are infinitely many q such that

$$q \, \| \, q\theta \, \| < 5^{-\frac{1}{2}}.$$

If θ is equivalent to $\frac{1}{2}(5^{\frac{1}{2}} - 1)$ the constant $5^{-\frac{1}{2}}$ cannot be replaced by any smaller number. If θ is not equivalent to $\frac{1}{2}(5^{\frac{1}{2}} - 1)$ there are infinitely many q such that

$$q \, \| \, q\theta \, \| < 2^{-\frac{3}{2}}.$$

Proof. We may confine our attention to best approximations. Write

$$A_n = q_n \| q_n \theta \|.$$

By Lemma 2, Corollary 3, we have $q_n \| q_{n-1}\theta \| + q_{n-1} \| q_n\theta \| = 1$, and so

$$\lambda^2 A_n - \lambda + A_{n-1} = 0, \quad \lambda = q_{n-1}/q_n. \tag{1}$$

Similarly $\quad \mu^2 A_n - \mu + A_{n+1} = 0, \quad \mu = q_{n+1}/q_n.$ $\tag{2}$

Here $\quad \mu - \lambda = (q_{n+1} - q_{n-1})/q_n = a_n.$ $\tag{3}$

We shall eliminate λ, μ from (1), (2), (3). On subtracting (2) from (1) and using (3),

$$a_n A_n(\lambda + \mu) = a_n + A_{n-1} - A_{n+1}. \tag{4}$$

On squaring (3) and (4) and adding,

$$2a_n^2 A_n^2(\lambda^2 + \mu^2) = a_n^4 A_n^2 + (a_n + A_{n-1} - A_{n+1})^2. \tag{5}$$

Finally, on adding (1) and (2) and using (4), (5) we have

$$a_n^2 A_n^2 + 2A_n(A_{n-1} + A_{n+1}) = 1 - a_n^{-2}(A_{n-1} - A_{n+1})^2 \leqslant 1. \tag{6}$$

The left-hand side of (6) is at least $(a_n^2 + 4) \min (A_{n-1}^2, A_n^2, A_{n+1}^2)$, and so either

$$\min (A_{n-1}, A_n, A_{n+1}) < 5^{-\frac{1}{2}}, \tag{7}$$

or $a_n = 1$, $A_{n-1} = A_n = A_{n+1} = 5^{-\frac{1}{2}}$. The second possibility cannot occur since (1) then gives the rational number $\lambda = q_{n-1}/q_n$ an irrational value $\frac{1}{2}(5^{\frac{1}{2}} \pm 1)$; and so (7) is always true.

If $a_n \geqslant 2$, we have

$$\min (A_{n-1}, A_n, A_{n+1}) < 2^{-\frac{3}{2}},$$

similarly. There are thus infinitely many solutions of

$$A_n < 2^{-\frac{3}{2}},$$

except, possibly, when

$$a_n = 1 \quad (\text{all } n \geqslant \text{ some } N).$$

These exceptional θ are, by Theorem IV, all equivalent to

$$\xi = [1, 1, 1, \ldots].$$

Since $\xi^{-1} = 1 + \xi$ and $0 < \xi < 1$, we have

$$\xi = \tfrac{1}{2}(5^{\frac{1}{2}} - 1).$$

It remains only to verify that if $\theta = \xi$ and $\kappa < 5^{-\frac{1}{2}}$ there are only a finite number of solutions of $q \, \| \, q\theta \, \| < \kappa$; and we may confine attention to best approximations p_n/q_n. But, by (2·15),

$$q_n \, \| \, q_n \theta \, \| = (1 + \theta_{n+1} + \phi_{n-1})^{-1},$$

where
$$\theta_{n+1} = [1, 1, 1, \ldots] = \xi,$$

and
$$\phi_{n-1} = \underbrace{[1, \ldots, 1]}_{n-1 \text{ digits}} \to \xi \quad (n \to \infty),$$

by Lemma 4. Hence

$$q_n \, \| \, q_n \theta \, \| \to (1 + 2\xi)^{-1} = 5^{-\frac{1}{2}},$$

as required. In the next chapter we shall prove much more than Theorem V by other means.

5. Simultaneous approximation. We sometimes wish to approximate a set of numbers $\theta_1, \ldots, \theta_n$ by fractions

$$p_1/q, \ldots, p_n/q$$

with a common denominator q; or, what is the same thing, to make $\| \, q\theta_1 \, \|, \ldots, \| \, q\theta_n \, \|$ simultaneously small. There is one quite general result:

THEOREM VI. *Let*

$$L_j(\mathbf{x}) = \sum_i \theta_{ji} x_i \quad (1 \leqslant i \leqslant m, \; 1 \leqslant j \leqslant n),$$

be n linear forms in m variables. To every real $X > 1$ there is an integral $\mathbf{x} \neq 0$ such that

$$\| \, L_j(\mathbf{x}) \, \| < X^{-m/n}, \quad |\, x_i \,| \leqslant X \quad (1 \leqslant i \leqslant m, 1 \leqslant j \leqslant n).$$

Proof. As in the second proof of Theorem I it is enough to find integers $x_1, \ldots, x_m, y_1, \ldots, y_n$ not all 0 such that

$$|\, L_j(\mathbf{x}) - y_j \,| < X^{-m/n} \quad (1 \leqslant j \leqslant n),$$

$$|\, x_i \,| \leqslant X \quad (1 \leqslant i \leqslant m).$$

The determinant of the $m + n$ forms in the $m + n$ variables on the left-hand side is 1, and as the product of the right-hand sides is 1 Theorem VI follows from Minkowski's linear forms theorem (Theorem III of Appendix B).

In particular, taking $m = 1$,

$$q^{1/n} \max (\| q\theta_1 \|, ..., \| q\theta_n \|) < 1,$$

for infinitely many integers $q > 0$. This may be improved slightly:

THEOREM VII. *There are infinitely many integer solutions of*

$$q^{1/n} \max (\| q\theta_1 \|, ..., \| q\theta_n \|) < n/(n+1).$$

Note. When $n = 1$ we have seen that $n/(n+1) = \frac{1}{2}$ on the right-hand side may be replaced by $5^{-\frac{1}{2}}$ but by nothing smaller. We do not know the best constants for $n > 1$.

Proof. Let $t > 1$ be arbitrary. By Theorem IV of Appendix B there are integers $x_1, ..., x_n, y$ not all 0 such that

$$t^{-n} | y | + t | \theta_j y - x_j | \leqslant (n+1)^{1/(n+1)} \quad (1 \leqslant j \leqslant n), \qquad (1)$$

since this $(n+1)$-dimensional region has volume 2^{n+1} (as may be verified by putting $z_j = t(\theta_j y - x_j) (1 \leqslant j \leqslant n); z_{n+1} = t^{-n} y$). Further, $y = 0$ implies $x_1 = x_2 = ... = x_n = 0$ if t is large enough, and thus we may assume $y > 0$. But then $y^{1/n} | \theta_j y - x_j | \leqslant n/(n+1)$ by the inequality of the arithmetic and geometric means applied to (1) with the left-hand side written as

$$t^{-n} y + \underbrace{n^{-1} t | \theta_j y - x_j | + ... + n^{-1} t | \theta_j y - x_j |}_{n \text{ summands}}.$$

Finally, infinitely many distinct solutions are obtained as $t \to \infty$, if at least one θ_j is irrational. If $\theta_1, ..., \theta_n$ are all rational there is an integer $Q > 0$ such that the $Q\theta_j$ are all integers. Then all positive multiples q of Q clearly satisfy the conditions of the theorem.

There is an analogue of Theorem VII for

$$u^n \| u_1 \theta_1 + ... + u_n \theta_n \|; \quad u = \max (| u_1 |, ..., | u_n |),$$

where $\theta_1, ..., \theta_n$ are given and $u_1, ..., u_n$ are integers not all 0.

Theorem VI is the best possible in a certain sense as the following theorem shows. The proof requires a certain knowledge of algebraic number theory: however, the results of the rest of the chapter are not required later in the book. For another proof, when $m = 1$ or $n = 1$, see Theorem III of Chapter V.

THEOREM VIII. *For any positive integers m, n there exists a constant $\gamma > 0$ and linear forms $L_j(\mathbf{x})$ $(1 \leqslant j \leqslant n)$ such that*

$$(\max_i |x_i|)^m (\max_j \|L_j(\mathbf{x})\|)^n \geqslant \gamma$$

for all integral $\mathbf{x} = (x_1, \dots, x_m) \neq \mathbf{0}$.

Proof. Put $l = m + n$ (> 1). There exist sets of real conjugate algebraic integers $\phi_1, \dots, \phi_l$ of degree l (see below). Write

$$Q_k(\mathbf{x}, \mathbf{y}) = \sum_{j=1}^{n} \phi_k^{j-1} y_j + \sum_{i=1}^{m} \phi_k^{n+i-1} x_i \quad (1 \leqslant k \leqslant l). \tag{2}$$

For integral $\mathbf{x}$, $\mathbf{y}$ not both $\mathbf{0}$ the $Q_k(\mathbf{x}, \mathbf{y})$ are conjugate algebraic integers not 0. In particular $\prod_k Q_k(\mathbf{x}, \mathbf{y})$ is then a rational integer not 0; and so

$$\prod_k |Q_k(\mathbf{x}, \mathbf{y})| \geqslant 1. \tag{3}$$

But now for $k = 1, \dots, n$ we may write (2) in the shape

$$Q_k(\mathbf{x}, \mathbf{y}) = \sum_{j=1}^{n} \phi_k^{j-1} (y_j - L_j(\mathbf{x})) \quad (k \leqslant n), \tag{4}$$

for some linear forms $L_j(\mathbf{x})$: indeed the $L_j(\mathbf{x})$ may be obtained by equating $Q_1, \dots, Q_n$ to 0 and solving for $y_1, \dots, y_n$ in terms of $\mathbf{x}$. For $k > n$ we then have

$$Q_k(\mathbf{x}, \mathbf{y}) = \sum_{j=1}^{n} \phi_k^{j-1} (y_j - L_j(\mathbf{x})) + \sum_{i=1}^{m} \omega_{ki} x_i \quad (k > n), \tag{5}$$

for some constants ω_{ki}.

Now let $\mathbf{x} \neq \mathbf{0}$ be integral and put

$$X = \max_i |x_i|, \quad C = \max_j \|L_j(\mathbf{x})\|.$$

Let $y_1, \dots, y_n$ be the integers such that

$$\|L_j(\mathbf{x})\| = |L_j(\mathbf{x}) - y_j|.$$

Then, by (4), $\qquad |Q_k(\mathbf{x}, \mathbf{y})| \leqslant \gamma_1 C \quad (k \leqslant n), \tag{6}$

where γ_1 (as $\gamma_2, \dots$ below) depends only on the ϕ_k but not on $\mathbf{x}$. Similarly, by (5),

$$|Q_k(\mathbf{x}, \mathbf{y})| \leqslant \gamma_2 C + \gamma_3 X \leqslant \gamma_4 X \quad (k > n), \tag{7}$$

since $C < 1 \leqslant X$.

From (6), (7) we have

$$\left| \prod_{k} Q_k(\mathbf{x}, \mathbf{y}) \right| \leqslant \gamma_1^n \gamma_4^m C^n X^m. \tag{8}$$

But now (1) with $\gamma = \gamma_1^{-n} \gamma_4^{-m}$ follows from (3) and (8).

[The ϕ_j may be taken as the roots of

$$(\phi - a_1 q) \dots (\phi - a_l q) - 1 = 0, \tag{9}$$

where $a_1, \dots, a_l$ are any distinct rational integers and q is sufficiently large. If q is large this has l real roots ϕ_k, where

$$0 < | \phi_k - a_k q | < K_1 q^{-l+1}, \tag{10}$$

and K_1 (as K_2 below) depends on $a_1, \dots, a_l$ but not on q. The ϕ_k are certainly algebraic integers. If they were not all conjugate, then, on rearranging $a_1, \dots, a_l$, we could suppose that

$$\phi_1, \dots, \phi_L (L < l)$$

are a set of conjugates; so $\prod_{\lambda \leqslant L} (a_1 q - \phi_\lambda)$ would be a rational integer. But, by (10),

$$0 < \left| \prod_{\lambda \leqslant L} (a_1 q - \phi_\lambda) \right| < K_2 q^{L-l} < 1,$$

if q is large enough; a contradiction.]

NOTES

§ 1. Another type of proof of Theorem I depends on Farey Section (HARDY & WRIGHT (1938), Chapter III).

§ 2. The continued fractions discussed are called the 'regular' continued fractions. They have two useful properties: (i) The sequence of a_n relating successive convergents is entirely arbitrary, (ii) The convergents p_n/q_n are characterized by a simple intrinsic property (that of being 'best approximations'). No other continued fraction algorithm has both (i) and (ii), e.g. the 'diagonal continued fractions' lack (i) and the 'continued fractions to the nearest integer' lack (ii).

The whole machinery can be carried over to the product $\xi\eta$ of two linear forms $\xi = \alpha x + \beta y$, $\eta = \gamma x + \delta y$ ($\alpha, \beta, \gamma, \delta$ real, x, y run through integer values). A pair of integers x_n, y_n gives a 'best approximation' if there is no solution in integers $(x, y) \neq (0, 0)$ of

$$| \xi(x, y) | < | \xi(x_n, y_n) |, \qquad | \eta(x, y) | < | \eta(x_n, y_n) |.$$

Many writers put p_{n-1}, q_{n-1} for our p_n, q_n. But our convention gives a greater symmetry between the forms ξ, η in the generalization just discussed; a symmetry which is reflected in the symmetry of (2·15) in θ_{n+1} and ϕ_{n-1}.

For two wider accounts of the theory from different aspects see KHINTCHINE (1935) and PERRON (1913) and for an extension to quadratic number-fields see POITOU (1953).

§ 4. The argument is Professor Davenport's. For an 'asymmetric' generalization see SEGRE (1945), BARNES & SWINNERTON-DYER (1955) and TORNHEIM (1955).

§ 5. For the present state of knowledge see DAVENPORT (1954).

CHAPTER II

THE MARKOFF CHAIN†

1. Introduction. As Markoff has shown, Theorem V of Chapter I is capable of extension. For all irrational θ the inequality

$$q\|q\theta\| < 5^{-\frac{1}{2}} \tag{1}$$

has infinitely many solutions. If θ is equivalent to $\frac{1}{2}(5^{\frac{1}{2}} - 1)$, i.e. to a root of

$$\theta^2 + \theta - 1 = 0, \tag{2}$$

the constant $5^{-\frac{1}{2}}$ cannot be improved. If not, there are infinitely many solutions of

$$q\|q\theta\| < 2^{-\frac{3}{2}}, \tag{3}$$

where the constant cannot be improved if θ is equivalent to a root of

$$\theta^2 + 2\theta - 1 = 0. \tag{4}$$

Otherwise, there are infinitely many solutions of

$$q\|q\theta\| < 5/(221)^{\frac{1}{2}}, \tag{5}$$

where the constant cannot be improved for θ equivalent to a root of

$$5\theta^2 + 11\theta - 5 = 0. \tag{6}$$

Otherwise, there are infinitely many solutions of

$$q\|q\theta\| < 13/(1517)^{\frac{1}{2}}, \tag{7}$$

where the constant cannot be improved for θ equivalent to a root of

$$13\theta^2 + 29\theta - 13 = 0. \tag{8}$$

And so on indefinitely. The sequence of numbers $5^{-\frac{1}{2}}, 2^{-\frac{3}{2}}, 5/(221)^{\frac{1}{2}}, 13/(1517)^{\frac{1}{2}}, \ldots$ tends to $\frac{1}{3}$.

There is a closely related chain of theorems relating to INDEFINITE QUADRATIC FORMS, that is, expressions of the shape

$$f(x, y) = \alpha x^2 + \beta xy + \gamma y^2, \tag{9}$$

† The results in this chapter are not used elsewhere. It might well be omitted at a first reading.

with two distinct real linear factors. The DISCRIMINANT

$$\delta(f) = \delta = \beta^2 - 4\alpha\gamma, \tag{10}$$

is then (strictly) positive. Two quadratic forms $f(x,y), f'(x,y)$ are EQUIVALENT if there are integers a, b, c, d such that

$$\left. \begin{array}{c} f'(ax+by, cx+dy) = f(x,y), \\ ad - bc = \pm 1, \end{array} \right\} \tag{11}$$

identically in x, y. Clearly the relation is symmetrical between f and f'. Further, it is readily verified that if f is equivalent to f' and f' to f'' then f is equivalent to f''; so that we have an equivalence relation in the usual sense. It is also easily verified that two equivalent forms have the same discriminant.

This equivalence is related to that for real numbers given earlier. If $f(\theta, 1) = 0$ and (11) holds, we have

$$f'(a\theta + b, c\theta + d) = f(\theta, 1) = 0,$$

i.e. $f'(\theta', 1) = 0$, where $\theta' = (a\theta + b)/(c\theta + d)$. Thus θ is equivalent to one of the roots of $f'(\theta', 1) = 0$.

We write†

$$\mu(f) = \inf |f(x,y)| \quad (x, y \text{ integers not both } 0).$$

Since by (11) two equivalent forms take precisely the same values as x, y run over all integer values, we have

$$\mu(f) = \mu(f') \quad (f' \text{ equivalent to } f).$$

Further, if $\lambda \neq 0$ is a real number we have

$$\mu(\lambda f) = |\lambda| \mu(f), \quad \delta(\lambda f) = \lambda^2 \delta(f);$$

so $\mu(f) \delta^{-\frac{1}{2}}(f)$ is unaffected either by replacing f by an equivalent form or by multiplying f by a constant.

The chain of theorems is now as follows:

$$\mu(f) \leqslant 5^{-\frac{1}{2}} \delta^{\frac{1}{2}}(f),$$

the sign of equality being required only for forms equivalent to a multiple of $x^2 + xy - y^2$. Otherwise

$$\mu(f) \leqslant 2^{-\frac{1}{2}} \delta^{\frac{1}{2}}(f),$$

† For the notation 'inf' see p. x.

with equality only for forms equivalent to a multiple of $x^2 + 2xy - y^2$. And so on. The numbers $5^{-\frac{1}{2}}$, $2^{-\frac{3}{2}}$, ... are the same as those occurring in the chain of theorems for approximations; and if the forms here are denoted by $f(x, y)$ then the θ in the approximation theorem are defined by $f(\theta, 1) = 0$.

The chain theorem for forms is much easier to prove if one assumes that $\mu(f)$ is ATTAINED, that is, that there are integers x_0, y_0 such that

$$|f(x_0, y_0)| = \mu(f).$$

Once the chain of theorems for forms has been proved for this special case we shall deduce the chain for forms in its full generality and the chain for approximations by a quite general technique ('isolation').

In § 2 we discuss the theory of quadratic forms and its relation to approximations. In particular we prove the theorem on which the isolation technique depends. In §§ 3, 4 we define and investigate the special quadratic forms which occur in the chains of theorems. Finally in §§ 5, 6 we enunciate and prove the two chains of theorems.

2. Indefinite binary quadratic forms. Throughout this section

$$f(x, y) = \alpha x^2 + \beta xy + \gamma y^2$$

is an indefinite quadratic form of discriminant

$$\delta = \beta^2 - 4\alpha\gamma > 0.$$

The roots of $f(x, 1) = 0$ are θ, ϕ; so that

$$f(x, y) = \alpha L(x, y) M(x, y),$$

where $\quad\quad L(x, y) = x - \theta y, \quad M(x, y) = x - \phi y$

and $\quad\quad\quad |\alpha(\theta - \phi)| = \delta^{\frac{1}{2}}.$

LEMMA 1. *Suppose that there are coprime integers a, b such that $f(a, b) = \alpha' \neq 0$. Then there are integers c, d with $ad - bc = 1$ for which*

$$f(ax + cy, bx + dy) = \alpha' x^2 + \beta' xy + \gamma' y^2$$

with $\quad\quad\quad\quad |\beta'| \leqslant |\alpha'|.$

Proof. Since a, b are coprime there are certainly integers c', d' with $ad' - bc' = 1$. Then

$$f(ax + c'y, bx + d'y) = \alpha'x^2 + \beta''xy + \gamma''y^2$$

for some β'', γ''. There is some integer n such that

$$|\beta'' + 2n\alpha'| \leqslant |\alpha'|.$$

Clearly $c = c' + na$, $d = d' + nb$ do what is required.

COROLLARY. *If $\alpha' > 0$ then $f(x, y)$ is also equivalent to a form $\alpha'x^2 + \beta'''xy + \gamma'''y^2$ with $2\alpha' \leqslant \beta''' \leqslant 3\alpha'$.*

Proof. Write $\alpha'x^2 + \beta'xy + \gamma'y^2 = f'(x,y)$. If $\beta' \geqslant 0$ take $f'(x+y, y)$ and if $\beta' < 0$ take $f'(x+y, -y)$.

LEMMA 2. ('Compactness lemma.') *Let*

$$f_j(x, y) = \alpha_j x^2 + \beta_j xy + \gamma_j y^2 \quad (1 \leqslant j < \infty).$$

Suppose that

$$0 < K_1 \leqslant |\alpha_j| \leqslant K_2, \quad |\beta_j| \leqslant K_3 |\alpha_j|,$$

for all sufficiently large j, where K_1, K_2, K_3 are independent of j. Suppose that
$$\lim (\beta_j^2 - 4\alpha_j\gamma_j) = \delta$$

exists. Then there is a subsequence $f_{j_s}(x, y)$ which tends to a limit $f(x, y) = \alpha x^2 + \beta xy + \gamma y^2$ in the sense that

$$\alpha_{j_s} \to \alpha, \quad \beta_{j_s} \to \beta, \quad \gamma_{j_s} \to \gamma.$$

Further, $\beta^2 - 4\alpha\gamma = \delta$.

Proof. The hypotheses imply, for large enough j, that

$$|\beta_j| \leqslant K_4, \quad |\gamma_j| \leqslant K_5,$$

where K_4, K_5 are independent of j. The points $P_j = (\alpha_j, \beta_j, \gamma_j)$ therefore lie in a bounded part of 3-dimensional Euclidean space. Hence there must be a subsequence P_{j_s} tending to a limit point $P = (\alpha, \beta, \gamma)$, say. Clearly α, β, γ do what is required.

COROLLARY. *If α_j, β_j, γ_j are integers then $f_j(x, y) = f(x, y)$ for infinitely many j.*

Proof. Clear.

LEMMA 3. *Let α, β, γ be rationals such that θ, ϕ are irrational (i.e. δ is not a perfect square). Then for some η with $0 < \eta < 1$ and integers a, b, c, d with $ad - bc = 1$ we have identically*

$$L(ax + by, cx + dy) = \eta L(x, y),$$
$$M(ax + by, cx + dy) = \eta^{-1} M(x, y).$$

Proof.† Without loss of generality α, β, γ are integers. For this proof only, write $\mathbf{x} = (x, y)$; and if $\mathbf{S} = \begin{pmatrix} a & c \\ b & d \end{pmatrix}$ is a 2×2 matrix write $\mathbf{xS} = (ax + by, cx + dy)$. The $\mathbf{S}$ with a, b, c, d integers and $ad - bc = 1$ form a group, i.e. if $\mathbf{S}_1, \mathbf{S}_2$ are of this shape so are $\mathbf{S}_1 \mathbf{S}_2$ and $\mathbf{S}_1^{-1}$.

By Theorem III of Appendix B, for any $\epsilon > 0$ there is an integral $\mathbf{x}^{(1)} = (x_1, y_1) \neq \mathbf{0}$ such that

$$|L(\mathbf{x}^{(1)})| < \epsilon, \quad |M(\mathbf{x}^{(1)})| \leqslant |\theta - \phi| \epsilon^{-1}.$$

Hence　　　$|f(\mathbf{x}^{(1)})| = |\alpha L(\mathbf{x}^{(1)}) M(\mathbf{x}^{(1)})| < |\alpha(\theta - \phi)|.$

Without loss of generality x_1, y_1 are coprime. But $L(\mathbf{x}^{(1)}) \neq 0$ since θ is irrational and so, on letting $\epsilon \to 0$, we have an infinite sequence of coprime integral vectors $\mathbf{x}^{(r)} = (x_r, y_r)$ with

$$|f(\mathbf{x}^{(r)})| < |\alpha(\theta - \phi)|, \ L(\mathbf{x}^{(r)}) \to 0.$$

By writing $-\mathbf{x}^{(r)}$ for $\mathbf{x}^{(r)}$ if necessary, we have

$$L(\mathbf{x}^{(r)}) > 0, \ L(\mathbf{x}^{(r)}) \to 0. \tag{1}$$

By Lemma 1 there exists an $\mathbf{S}_r = \begin{pmatrix} x_r & y_r \\ z_r & t_r \end{pmatrix}$ with integral z_r, t_r and $x_r t_r - z_r y_r = 1$ such that

$$f(\mathbf{xS}_r) = \alpha_r x^2 + \beta_r xy + \gamma_r y^2, \quad \alpha_r = f(\mathbf{x}^{(r)}),$$

with $|\beta_r| \leqslant |\alpha_r|$, $\beta_r^2 - 4\alpha_r \gamma_r = \beta^2 - 4\alpha\gamma$ and $1 \leqslant |\alpha_r| < |\alpha(\theta - \phi)|$, since $f(\mathbf{x}^{(r)})$ is a non-zero integer. By Lemma 2, Corollary, we may suppose, on taking a subsequence of the $\mathbf{S}_r$ instead of $\mathbf{S}_r$, that

$$f(\mathbf{xS}_r) = \phi(\mathbf{x}) \quad \text{(say)}, \tag{2}$$

---

† In the application to the Markoff chain we can always write down a, b, c, d explicitly. We give the general existence theorem so as to be able to express Lemma 4 and Theorem I in complete generality.

is independent of r. Let $\phi(\mathbf{x}) = \lambda(\mathbf{x})\,\mu(\mathbf{x})$ be any factorization into linear factors. Another is

$$\phi(\mathbf{x}) = f(\mathbf{x}\mathbf{S}_r) = \alpha L(\mathbf{x}\mathbf{S}_r)\,M(\mathbf{x}\mathbf{S}_r).$$

Hence for each r, identically in $\mathbf{x}$,

either $\qquad L(\mathbf{x}\mathbf{S}_r) = \nu_r \lambda(\mathbf{x}), \quad M(\mathbf{x}\mathbf{S}_r) = \pi_r \mu(\mathbf{x}),$ $\qquad$ (3)

or $\qquad L(\mathbf{x}\mathbf{S}_r) = \nu_r \mu(\mathbf{x}), \quad M(\mathbf{x}\mathbf{S}_r) = \pi_r \lambda(\mathbf{x}),$

for some real ν_r, π_r. By taking a subsequence again we may suppose that always the same alternative occurs, and, by interchanging $\lambda(\mathbf{x})$, $\mu(\mathbf{x})$ if need be, that it is (3).

But $(1, 0)\,\mathbf{S}_r = \mathbf{x}^{(r)}$ by construction. Hence on putting $\mathbf{x} = (1, 0)$ in (3) and using (1) we have

$$0 < \nu_r/\nu_1 = L(\mathbf{x}^{(r)})/L(\mathbf{x}^{(1)}) \to 0 \quad (r \to \infty).$$

Put $\eta = \nu_r/\nu_1$, $\mathbf{T} = \mathbf{S}_1^{-1}\mathbf{S}_r$, where r is so large that $0 < \eta < 1$. Then

$$f(\mathbf{x}\mathbf{T}) = \phi(\mathbf{x}\mathbf{S}_1^{-1}) = f(\mathbf{x}),$$ $\qquad$ (4)

by (2) with $\mathbf{x}\mathbf{S}_1^{-1}$ for $\mathbf{x}$. Similarly (3) gives

$$L(\mathbf{x}\mathbf{T}) = \nu_r \lambda(\mathbf{x}\mathbf{S}_1^{-1}) = \eta L(\mathbf{x}).$$

Finally $\qquad\qquad M(\mathbf{x}\mathbf{T}) = \eta^{-1}M(\mathbf{x})$

by (4), since $f(\mathbf{x}\mathbf{T}) = \alpha L(\mathbf{x}\mathbf{T})\,M(\mathbf{x}\mathbf{T})$. This proves the lemma with $\mathbf{T} = \begin{pmatrix} a & c \\ b & d \end{pmatrix}$.

COROLLARY. *Let x_0, y_0, n be any integers. Then there are integers x_1, y_1 such that $f(x_1, y_1) = f(x_0, y_0)$ and*

$$L(x_1, y_1) = \eta^n L(x_0, y_0), \quad M(x_1, y_1) = \eta^{-n}M(x_0, y_0).$$

Proof. For $n > 0$ and the $\mathbf{T}$ of the proof we have

$$L(\mathbf{x}\mathbf{T}^n) = \eta L(\mathbf{x}\mathbf{T}^{n-1}) = \ldots = \eta^n L(\mathbf{x}),$$
$$M(\mathbf{x}\mathbf{T}^n) = \ldots \qquad\qquad = \eta^{-n}M(\mathbf{x}).$$

On writing $\mathbf{x}\mathbf{T}^{-1}$ for $\mathbf{x}$ in $L(\mathbf{x}\mathbf{T}) = \eta L(\mathbf{x})$ we have

$$L(\mathbf{x}\mathbf{T}^{-1}) = \eta^{-1}L(\mathbf{x}),$$

and so, similarly,

$$L(\mathbf{x}\mathbf{T}^n) = \eta^n L(\mathbf{x}), \quad M(\mathbf{x}\mathbf{T}^n) = \eta^{-n}M(\mathbf{x}) \quad \text{for } n < 0.$$

The corollary follows on putting $(x_1, y_1) = (x_0, y_0)\,\mathbf{T}^n$.

LEMMA 4. *Suppose that θ is irrational and is a root of*

$$f(\theta, 1) = 0.$$

As before, put $\qquad \nu = \nu(\theta) = \liminf q \parallel q\theta \parallel,$
and

$$\mu = \mu(f) = \inf |f(x,y)|, \quad (x, y \text{ integers not both } 0).$$

Then:

A. $\qquad\qquad\qquad \nu(\theta) \geqslant \delta^{-\frac{1}{2}} \mu(f),$ \hfill (5)

whatever the values of α, β, γ (rational or irrational).

B. *If α, β, γ are rational, there is always equality in* (5) *and $\mu(f)$ is attained.*

C. *If, in addition, $f(x,y)$ takes both values $\pm\mu$ for integer values of the variables, there are infinitely many integers q such that*

$$q \parallel q\theta \parallel < \nu.$$

Proof. The proofs repose on the obvious identity

$$f(p, q) = \alpha(p - \theta q)(p - \phi q),$$
$$= \alpha(\theta - \phi) q(p - \theta q) + \alpha(p - \theta q)^2. \hfill (6)$$

Suppose first that ν' is any number $> \nu$, so there are certainly solutions of $q\,|\,q\theta - p\,| < \nu'$ with q arbitrarily large. Then (6) gives

$$|f(p, q)| \leqslant |\alpha|\,|\theta - \phi|\,\nu' + |\alpha|\,\nu'^2 q^{-2}.$$

But $\qquad\qquad |f(p, q)| \geqslant \mu \quad$ and $\quad |\alpha(\theta - \phi)| = \delta^{\frac{1}{2}},$

so $\qquad\qquad\qquad \mu \leqslant \nu'\delta^{\frac{1}{2}} + |\alpha|\,\nu'^2 q^{-2}.$

Hence $\mu \leqslant \nu\delta^{\frac{1}{2}}$, since q is arbitrarily large and ν' is any number larger than ν. This proves A.

If α, β, γ are rational there is an integer h such that $hf(x,y)$ is always an integer for integers x, y. Thus $|hf(x,y)|$ must attain its lower bound, that is, $|f(p, q)| = \mu$ for integers p, q. By Lemma 3, Corollary, there are such p, q with $|q\theta - p|$ arbitrarily small. The result $\mu \geqslant \nu\delta^{\frac{1}{2}}$, and so B, follows on reversing the previous argument.

Finally, if $f(x,y)$ takes both values $\pm\mu$ there are integers $q > 0, p$ such that $f(p, q)$ has either value $\pm\mu$ and $|q\theta - p|$ is arbitrarily small. Then, for appropriate choice of sign,

$$\mu - |f(p, q)| > |\alpha(\theta - \phi) q(\theta q - p)|,$$

since the second term on the right-hand side of (6) has always the same sign as α and is smaller in absolute value than $|f(p,q)| = \mu$ if $|q\theta - p|$ is small enough. This proves C.

THEOREM I. ('Isolation theorem', Remak, Rogers.) *Suppose that* $f(x,y) = \alpha x^2 + \beta xy + \gamma y^2$, *where* α, β, γ *are rational but the roots* θ, ϕ *of* $f(x,1) = 0$ *are irrational. Let* $\mu > 0$ *be the minimum of* $|f(x,y)|$ *for integers* x, y *not both* 0. *Suppose, further, that both*

$$f(x,y) = \pm \mu \tag{7}$$

are soluble in integers. Then there is a $\mu' < \mu$ *and an* $\epsilon_0 > 0$ *depending only on* α, β, γ *with the following property*:
Let

$$f^*(x,y) = \alpha^*(x - \theta^* y)(x - \phi^* y) \tag{8}$$

be any quadratic form for which

$$|\alpha - \alpha^*| < \epsilon_0, \quad |\theta - \theta^*| < \epsilon_0, \quad |\phi - \phi^*| < \epsilon_0. \tag{9}$$

Then there are integers x_0, y_0 *not both* 0 *such that*

$$|f^*(x_0, y_0)| < \mu', \tag{10}$$

provided that f^* *is not of the shape* λf *for a constant* λ.

Note 1. Substantially, the theorem says that all forms 'near enough' to f, other than multiples of f, have a somewhat lower minimum. The last proviso is essential since the minimum of λf is $|\lambda| \mu$, and λ could be arbitrarily near 1.

Note 2. The real restriction is that both $f(x,y) = \pm \mu$ must be soluble. Thus the theorem does not apply to $x^2 - 3y^2$.

Proof. Without loss of generality α, β, γ are integers. If $\theta^* = \theta$, $\phi^* = \phi$, then f^* is a multiple of f. Hence, we may suppose, by symmetry, that

$$\theta^* \neq \theta, \quad \alpha > 0. \tag{11}$$

By hypothesis there exist integers x_1, y_1, x_2, y_2 such that

$$f(x_1, y_1) = \mu, \quad f(x_2, y_2) = -\mu,$$

and, after changes of sign if necessary, we may suppose, in the notation introduced on p. 20, that

$$L_1 = L(x_1, y_1) > 0, \quad M_1 = M(x_1, y_1) > 0, \quad \alpha L_1 M_1 = \mu; \tag{12}$$
$$L_2 = L(x_2, y_2) > 0, \quad M_2 = M(x_2, y_2) < 0, \quad \alpha L_2 M_2 = -\mu. \tag{13}$$

Denote by $c_1, c_2, \ldots$ positive constants depending only on $\alpha, \beta, \gamma, L_1, M_1, L_2, M_2, \eta$, where η is as in Lemma 3; i.e. essentially only on α, β, γ. Let μ' be any number such that

$$\mu > \mu' > \mu(1 - \eta^2). \tag{14}$$

It will be enough to show that the required x_0, y_0 exist provided that ϵ_0 is smaller than some number depending only on α, β, γ and μ'.

We write
$$L^*(x, y) = x - \theta^* y, \quad M^* = x - \phi^* y, \tag{15}$$

so that
$$L^* = (1 - \rho) L + \rho M, \quad M^* = \sigma L + (1 - \sigma) M, \tag{16}$$

where
$$\rho = \frac{\theta^* - \theta}{\phi - \theta}, \quad \sigma = \frac{\phi^* - \phi}{\theta - \phi}. \tag{17}$$

In particular
$$0 < |\rho| < c_1 \epsilon_0, \quad |\sigma| < c_1 \epsilon_0. \tag{18}$$

First case, $\rho \leqslant 0$. We determine the integer n by

$$\eta^{2n} \geqslant \frac{-\rho M_1}{(1 - \rho) L_1} > \eta^{2(n+1)}, \tag{19}$$

so that
$$0 < \eta^{2n} < c_2 \epsilon_0. \tag{20}$$

Now choose integers x_0, y_0, as we may by Lemma 3, Corollary, so that
$$L_0 = L(x_0, y_0) = \eta^n L_1, \quad M_0 = M(x_0, y_0) = \eta^{-n} M_1. \tag{21}$$

The corresponding values $L_0^* = L^*(x_0, y_0)$, $M_0^* = M^*(x_0, y_0)$ are given by (16). Hence by (19), (20), (21) they satisfy

$$0 \leqslant \eta^{-n} L_0^* = (1 - \rho) L_1 + \rho \eta^{-2n} M_1 \leqslant (1 - \rho)(1 - \eta^2) L_1$$
$$\leqslant (1 + c_1 \epsilon_0)(1 - \eta^2) L_1, \tag{22}$$

and
$$|\eta^n M_0^*| \leqslant |\sigma| \eta^{2n} |L_1| + (1 + |\sigma|) |M_1|$$
$$\leqslant c_3 \epsilon_0^2 + (1 + c_1 \epsilon_0) |M_1|$$
$$\leqslant (1 + c_4 \epsilon_0) |M_1|. \tag{23}$$

Then by (9), (11), (22), (23),

$$|f^*(x_0, y_0)| = |\alpha^* L_0^* M_0^*|$$
$$\leqslant (\alpha + \epsilon_0)(1 + c_1 \epsilon_0)(1 + c_4 \epsilon_0)(1 - \eta^2) |L_1 M_1|$$
$$\leqslant (1 + c_5 \epsilon_0)(1 - \eta^2) \alpha |L_1 M_1|$$
$$= (1 + c_5 \epsilon_0)(1 - \eta^2) \mu < \mu',$$

provided that ϵ_0 is small enough.

Second case, $\rho > 0$. Similar, except that L_2, M_2 take over the role of L_1, M_1.

COROLLARY. *If $\theta \neq \theta^*$ and if ϵ_0 is small enough we may suppose further that*
$$|x_0 - \theta^* y_0| < 1.$$

Proof. From (20), (21) and (22) we have
$$|x_0 - \theta^* y_0| = |L_0^*| < \eta^n (1 + c_1 \epsilon_0)(1 - \eta^2) L_1$$
$$< c_6 \eta^n < c_7 \epsilon_0^{\frac{1}{3}} < 1.$$

3. A Diophantine equation.

We first discuss the solution of
$$m^2 + m_1^2 + m_2^2 = 3m m_1 m_2 \tag{1}$$
in positive integers. The m which may occur are called the MARKOFF NUMBERS.

Suppose, first, that two of m, m_1, m_2 are equal, say $m_1 = m_2$. Then $m_1^2 \mid m^2$, say $m = d m_1$, and then $d^2 + 2 = 3 m_1 d$. Thus $d \mid 2$ and $d = 1$ or 2. In either case $m_1 = 1$ and we have the SINGULAR SOLUTIONS $(1, 1, 1)$ and $(2, 1, 1)$ (with their permutations). Otherwise m, m_1, m_2 are distinct.

The quadratic
$$\Phi(x) = x^2 - 3x m_1 m_2 + m_1^2 + m_2^2,$$
has the positive integer root m. The other root m', which satisfies $m + m' = 3 m_1 m_2$, $m m' = m_1^2 + m_2^2$, must also be a positive integer. If, say, $m_1 > m_2$ then we have
$$(m_1 - m)(m_1 - m') = \Phi(m_1) = 2m_1^2 + m_2^2 - 3m_1^2 m_2 < 0.$$

Hence, $\max(m_1, m_2)$ lies strictly between m and m' except (possibly) for the singular solutions. Thus every non-singular solution gives rise to three distinct solutions
$$(m', m_1, m_2), \quad (m, m_1', m_2), \quad (m, m_1, m_2'),$$
where
$$m' = 3m_1 m_2 - m, \quad m_1' = 3m m_2 - m_1, \quad m_2' = 3m m_1 - m_2. \tag{2}$$

We shall say that these three solutions are NEIGHBOURS of the original one. By taking neighbours successively we may hope to generate infinitely many solutions from a given one. If (m, m_1, m_2) is non-singular and

$$m = \max(m, m_1, m_2), \tag{3}$$

then
$$\left.\begin{aligned} m' &< \max(m_1, m_2) < m, \\ m_1' &> \max(m, m_2) = m, \quad m_2' > m. \end{aligned}\right\} \tag{4}$$

Thus one neighbour has a lower maximum element but two have a higher maximum. If we start with an arbitrary solution and take successively the neighbour with a lower maximum element we must ultimately come to a halt; and this can only be by reaching a singular solution. On the other hand $(1, 1, 1)$ has only the neighbour $(2, 1, 1)$ which has only the other neighbour $(5, 2, 1)$ (apart from permutations), as is easily verified. The solutions are thus arranged as in Figure 1. To sum up:

LEMMA 5. *All solutions may be derived through a chain of neighbours from* $(1, 1, 1)$. *Further*

$$\text{g.c.d.}\,(m, m_1) = \text{g.c.d.}\,(m, m_2) = \text{g.c.d.}\,(m_1, m_2) = 1.$$

Proof. Only the last statement needs verification. If, say, $d = \text{g.c.d.}\,(m, m_1)$ then $d \mid m_2$ from (1). Hence $d \mid m'$, $d \mid m_1'$, $d \mid m_2'$ by (2) and d divides the g.c.d. of the elements of any neighbouring solution. On working back to $(1, 1, 1)$ we have

$$d \mid \text{g.c.d.}\,(1, 1, 1).$$

From now on we shall always suppose that (3) (and hence (4)) holds for a non-singular solution. By (1) and since m, m_1, m_2 are coprime we may find integers k, k_1, k_2 such that†

$$\left.\begin{aligned} k &\equiv \frac{m_2}{m_1} \equiv \frac{-m_1}{m_2}\ (m), \quad 0 \leqslant k < m, \\ k_1 &\equiv \frac{m}{m_2} \equiv \frac{-m_2}{m}\ (m_1), \quad 0 \leqslant k_1 < m_1, \\ k_2 &\equiv \frac{m_1}{m} \equiv \frac{-m}{m_1}\ (m_2), \quad 0 < k_2 \leqslant m_2, \end{aligned}\right\} \tag{5}$$

† The $\leqslant$ signs have been placed to ensure the truth of Lemma 7. Here $a \equiv b/c\ (m)$ for integers a, b, c with c prime to m means $m \mid (ac - b)$.

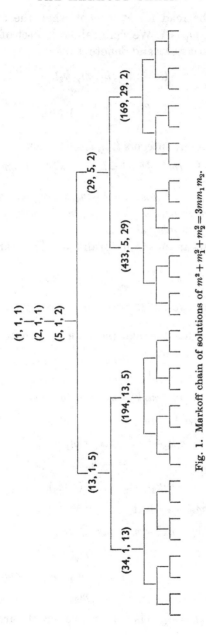

Fig. 1. Markoff chain of solutions of $m^2 + m_1^2 + m_2^2 = 3mm_1m_2$.

where $<$ can be read for $\leqslant$ except when the corresponding modulus m, m_1, m_2 is 1. We shall call such a set of members an ORDERED MARKOFF SET and denote it by

$$(m, k: m_1, k_1; m_2, k_2).$$

We note that

$$k^2 \equiv \frac{m_2}{m_1} \cdot \frac{-m_1}{m_2} \equiv -1 \ (m), \tag{6}$$

etc., and so there are integers l, l_1, l_2 such that

$$k^2 + 1 = lm, \quad k_1^2 + 1 = l_1 m_1, \quad k_2^2 + 1 = l_2 m_2. \tag{7}$$

LEMMA 6. *If $m > 1$ and $(m, k: m_1, k_1; m_2, k_2)$ is an ordered Markoff set, then so are $(m_1', k_1': m, k; m_2, k_2)$ and $(m_2', k_2': m_1, k_1; m, k)$ if k_1', k_2' are suitably chosen.*

Proof. Follows at once from (2) and (5). For example, by (5_1),

$$k \equiv \frac{m_2}{m_1} \equiv \frac{-m_2}{m_1'} \equiv \frac{-m_2'}{m_1} \ (m),$$

which is the analogue of (5_3) for $(m_2', k_2': m_1, k_1; m, k)$.

LEMMA 7. *For non-singular (m, m_1, m_2), we have*
$$mk_2 - m_2 k = m_1,$$
$$m_1 k - mk_1 = m_2,$$
$$m_1 k_2 - m_2 k_1 = m' = 3m_1 m_2 - m.$$

Proof. By (5) we have

$$mk_2 - m_2 k \equiv mk_2 \equiv m_1 \ (m_2)$$

and

$$\equiv -m_2 k \equiv m_1 \ (m),$$

and so

$$mk_2 - m_2 k \equiv m_1 \ (mm_2),$$

since g.c.d. $(m, m_2) = 1$. But

$$mk_2 - m_2 k - m_1 < mk_2 \leqslant mm_2$$

and

$$\geqslant m - m_2(m-1) - m_1$$
$$= (m + m_2 - m_1) - mm_2$$
$$> -mm_2.$$

Hence $mk_2 - m_2 k = m_1$. The other two proofs are similar (cf. Lemma 6).

4. The Markoff forms.

Let m, m_1, m_2 be positive integers with

$$m^2 + m_1^2 + m_2^2 = 3mm_1m_2, \quad m \geqslant \max(m_1, m_2). \tag{1}$$

As in § 3 denote by k the integer for which

$$m_1 k \equiv m_2 \ (m), \quad 0 \leqslant k < m, \tag{2}$$

and by l the integer defined by

$$k^2 + 1 = lm. \tag{2'}$$

The form F_m, defined by

$$mF_m(x, y) = mx^2 + (3m - 2k)xy + (l - 3k)y^2, \tag{3}$$

is called a MARKOFF FORM. In this section we retain the notation of § 3.

Identically $\qquad m^2 F_m(x, y) = \phi_m(y, z),$ $\qquad\qquad$ (4)

where $\qquad\qquad\qquad z = mx - ky,$ $\qquad\qquad\qquad$ (5)

and $\qquad\qquad\quad \phi_m(y, z) = y^2 + 3myz + z^2.$ $\qquad\qquad$ (6)

Trivially

$$\phi_m(y, z) = \phi_m(z, y) = \phi_m(-z, y + 3mz)$$
$$= \phi_m(z + 3my, -y). \tag{7}$$

By (5), (6) the discriminant of mF_m is $9m^2 - 4$, and so

$$F_m = \left(x + \frac{3m - 2k}{2m} y\right)^2 - \left(\frac{9}{4} - \frac{1}{m^2}\right) y^2. \tag{8}$$

The definition of F_m given is asymmetric in m_1, m_2. Suppose that $m_2 k' \equiv m_1 \ (m)$, $0 \leqslant k' < m$ and $k'^2 + 1 = l'm$. Let F_m' be the corresponding form. By (3·5) we have $k + k' \equiv 0 \ (m)$ and so either $m = 1$, $k = k' = 0$ or $m > 1$ and $k + k' = m$. In the first case $F_m' = F_m$ and in the second $F_m'(x, y) = F_m(x + 2y, -y)$ by (8). Since we deal only with equivalence of forms we need not consider F_m and F_m' separately. If we order m_1, m_2 so that $k \leqslant k'$ then

$$0 \leqslant 2k \leqslant m. \tag{9}$$

With this convention the first form is $x^2 + 3xy + y^2$, which is equivalent to $x^2 + xy - y^2$ given in the introduction.

To the genealogical tree of solutions of

$$m^2 + m_1^2 + m_2^2 = 3mm_1m_2$$

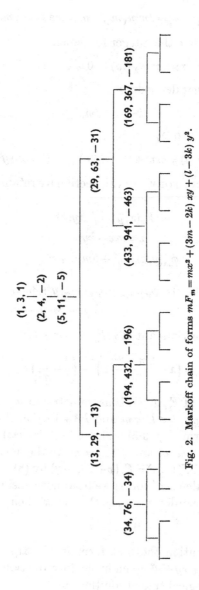

Fig. 2. Markoff chain of forms $mF_m = mx^2 + (3m - 2k)\,xy + (l - 3k)\,y^2$.

there corresponds a genealogical tree of Markoff forms (Figure 2). There is a slight ambiguity in the notation F_m since no one has shown that there cannot be two distinct solutions (m, m_1, m_2), (m, m_1^*, m_2^*) occurring in different portions of the tree. No case of this is known, and it seems improbable. However, there is no ambiguity in practice.

LEMMA 8. *For non-singular* (m, m_1, m_2),

$$F_m(k, m) = F_m(k - 3m, m) = 1,$$
$$F_m(k_1, m_1) = F_m(k_2 - 3m_2, m_2) = -1.$$

Proof.
$$m^2 F_m(k, m) = \phi_m(m, 0) = m^2$$

by (4), (5). Similarly (4), (5), (7) give

$$m^2 F_m(k - 3m, m) = \phi_m(m, -3m^2) = \phi_m(0, -m) = m^2.$$

By Lemma 7, $(x, y) = (k_1, m_1)$ gives $z = -m_2$, so

$$m^2 F_m(k_1, m_1) = \phi_m(m_1, -m_2)$$
$$= m_1^2 - 3mm_1 m_2 + m_2^2 = -m^2$$

by (1). Finally,

$$m^2 F_m(k_2 - 3m_2, m_2) = \phi_m(m_2, m_1 - 3mm_2)$$
$$= \phi_m(m_1, -m_2) = -m^2.$$

COROLLARY. *Let* $f(x, y) = x^2 + \beta xy + \gamma y^2$ *for some* β, γ *and suppose that*
$$f(k, m) \geqslant 1, \qquad f(k - 3m, m) \geqslant 1,$$
$$f(k_1, m_1) \leqslant -1, \quad f(k_2 - 3m_2, m_2) \leqslant -1.$$

Then $f(x, y) = F_m(x, y)$.
Proof. Let $F_m(x, y) = x^2 + \beta_m xy + \gamma_m y^2$. By the lemma, the four inequalities have the shape

$$B_1 \beta + C_1 \gamma \geqslant B_1 \beta_m + C_1 \gamma_m, \tag{10}$$
$$-B_2 \beta + C_2 \gamma \geqslant -B_2 \beta_m + C_2 \gamma_m, \tag{11}$$
$$-B_3 \beta - C_3 \gamma \geqslant -B_3 \beta_m - C_3 \gamma_m, \tag{12}$$
$$B_4 \beta - C_4 \gamma \geqslant B_4 \beta_m - C_4 \gamma_m, \tag{13}$$

where each B_j, C_j is positive (e.g. $B_1 = km$, $C_1 = m^2$, etc.). But then from (10), (11) we have $(B_2 C_1 + B_1 C_2)\gamma \geqslant (B_2 C_1 + B_1 C_2)\gamma_m$, i.e. $\gamma \geqslant \gamma_m$. Similarly $\gamma \leqslant \gamma_m$, $\beta \geqslant \beta_m$, $\beta \leqslant \beta_m$ and so $\beta = \beta_m$, $\gamma = \gamma_m$ as required.

LEMMA 9. $F_m(x, y)$ is equivalent to $-F_m(x, y)$.

Proof. This is true for

$$F_1(x, y) = x^2 + 3xy + y^2 \quad \text{and} \quad F_2(x, y) = x^2 + 2xy - y^2,$$

since

$$F_1(x + 2y, -x - y) = -F_1(x, y) \quad \text{and} \quad F_2(y, -x) = -F_2(x, y).$$

We may thus assume that (m, m_1, m_2) is non-singular and shall show that

$$F_m(k_1 x - l_1 y, m_1 x - k_1 y) = -F_m(x, y). \tag{14}$$

By Lemma 8 the two sides of (14) are equal when $(x, y) = (1, 0)$; and also when $(x, y) = (k_1, m_1)$ since then

$$k_1 x - l_1 y = k_1^2 - l_1 m_1 = 1 \quad \text{and} \quad m_1 x - k_1 y = 0.$$

The sides also† coincide for $(x, y) = (k, m)$ since then, by repeated use of Lemma 7,

$$m_1 x - k_1 y = m_1 k - k_1 m = m_2,$$

and

$$\begin{aligned}
m_1(k_1 x - l_1 y) &= m_1(k_1 k - l_1 m) \\
&= m_1 k_1 k - m(k_1^2 + 1) \\
&= k_1 m_2 - m = m_1 k_2 - 3 m_1 m_2 \\
&= m_1(k_2 - 3 m_2).
\end{aligned}$$

Hence (14) considered as a quadratic equation for y/x has three distinct roots; so it must be an identity.

COROLLARY. The roots θ, ϕ of $F_m(x, 1) = 0$ are equivalent.

Proof. By (14),

$$F_m(-k_1 \theta + l_1, -m_1 \theta + k_1) = -F_m(\theta, 1) = 0,$$

---

† Since the discriminants of the two forms (14) are equal we already have three independent functions of the three coefficients which are equal for the two forms. Unfortunately, since one of these is a quadratic we cannot deduce the identity. Hence the need for a third set of values.

i.e. $\qquad\qquad F_m(\theta', 1) = 0, \quad \theta' = \dfrac{-k_1\theta + l_1}{-m_1\theta + k_1}.$

If $\theta' = \theta$ then $m_1\theta^2 - 2k_1\theta + l_1 = 0$, which is impossible since $k_1^2 - m_1 l_1 = -1 < 0$. Hence $\theta' = \phi$.

LEMMA 10. $|F_m(x,y)| \geqslant 1$ *for all integers* $(x,y) \neq (0,0)$.

Proof. Let μ be the minimum of $|F_m(x,y)|$ for all integers $(x,y) \neq (0,0)$. Since mF_m has integer coefficients there are integers x_0, y_0 with $|F_m(x_0, y_0)| = \mu$. By Lemma 9 we may suppose that
$$F_m(x_0, y_0) = \mu \geqslant 0. \tag{15}$$

We use (4)–(7). A necessary and sufficient condition that integers (y,z) arise from an integral (x,y) in (5) is
$$z \equiv -ky \ (m). \tag{16}$$

We choose a solution of (15), if there is more than one, for which
$$|y_0| + |z_0| \text{ is a minimum}, \tag{17}$$
where $\qquad\qquad z_0 = mx_0 - ky_0.$

Suppose first, if possible, that
$$y_0 z_0 < 0, \quad |z_0| > |y_0|. \tag{18}$$

Put $y_1 = 3my_0 + z_0$, $z_1 = -y_0$, so (y_1, z_1) satisfies (16) since (y_0, z_0) does and $k^2 = lm - 1 \equiv -1 \ (m)$. Then
$$0 \leqslant m^2\mu = \phi_m(y_0, z_0) = \phi_m(y_1, z_1) = z_0 y_1 + y_0^2. \tag{19}$$

Hence, $-y_0^2 \leqslant z_0 y_1 < z_0^2$, the second inequality following trivially from (18); so, by (18) again, $|y_1| < |z_0|$. We should thus have $|y_1| + |z_1| < |y_0| + |z_0|$, a contradiction to (17). Similarly, $y_0 z_0 < 0$, $|y_0| > |z_0|$ leads to a contradiction. Since
$$\phi_m(y_0, -y_0) = -(3m-2)y_0^2 < 0,$$
we must have $\qquad\qquad y_0 z_0 \geqslant 0. \tag{20}$

But now $y_2 = z_0$, $z_2 = -y_0$ satisfy (16), as before, and
$$|\phi_m(y_2, z_2)| = |y_0^2 + z_0^2 - 3my_0 z_0|$$
$$\leqslant y_0^2 + z_0^2 + 3my_0 z_0 = m^2\mu. \tag{21}$$

There must be equality in (21) by the definition of μ. Hence $y_0 = 0$ or $z_0 = 0$. If $y_0 = 0$, then $m^2 \mu = \phi_m(0, z_0) = z_0^2 \geqslant m^2$ since $m \mid z_0$ by (16). Similarly $m^2 \mu = y_0^2 \geqslant m^2$ if $z_0 = 0$.

LEMMA 11. *Let*
$$f(x, y) = x^2 + \beta xy + \gamma y^2.$$
Suppose that
$$f(k_1, m_1) \leqslant -1, \quad f(k_2 - 3m_2, m_2) \leqslant -1.$$
Then $\beta^2 - 4\gamma \geqslant 4\Delta_m = 9 - 4m^{-2}$.

   Proof. Write
$$f(x, y) = (x + \tfrac{1}{2}\beta y)^2 - \Delta y^2,$$
$$F_m(x, y) = (x + \tfrac{1}{2}\beta_m y)^2 - \Delta_m y^2.$$
We must prove $\qquad\qquad \Delta \geqslant \Delta_m.$ (22)

By Lemma 8, $\qquad\qquad f(x, y) \leqslant F_m(x, y),$ (23)

both for $(x, y) = (k_1, m_1)$ and $(k_2 - 3m_2, m_2)$. We may write (23) as

$$\Delta - \Delta_m \geqslant \left(\frac{x}{y} + \frac{\beta}{2}\right)^2 - \left(\frac{x}{y} + \frac{\beta_m}{2}\right)^2. \qquad (24)$$

If $\beta \geqslant \beta_m$ then (22) follows from (24) with $(x, y) = (k_1, m_1)$ since $k_1/m_1 \geqslant 0$. If $\beta \leqslant \beta_m$, then (22) follows from (24) with
$$(x, y) = (k_2 - 3m_2, m_2),$$
since
$$\frac{3m_2 - k_2}{m_2} \geqslant 2 > \frac{3m - 2k}{2m} = \frac{\beta_m}{2}.$$

LEMMA 12. *Let*
$$f(x, y) = x^2 + \beta xy + \gamma y^2.$$
Suppose that
$$f(k, m) \leqslant -1, \quad f(k - 3m, m) \leqslant -1.$$
Then $\qquad\qquad \beta^2 - 4\gamma \geqslant 9 + 4m^{-2} > 9.$

   Proof. With the notation of the previous proof,
$$f(x, y) \leqslant F_m(x, y) - 2$$
both for $(x, y) = (k, m)$ and $(k - 3m, m)$, by Lemma 8. The rest of the argument is similar.

LEMMA 13. *Let*
$$f(x, y) = x^2 + \beta xy + \gamma y^2,$$
where $\qquad 2 \leqslant \beta \leqslant 3 \quad and \quad 0 < \beta^2 - 4\gamma < 9.$

Suppose that $|f(x, y)| \geqslant 1$ *for all integers* $(x, y) \neq (0, 0)$. *Then* $f(x, y)$ *is a* $F_m(x, y)$.

Proof. For all integers $(x, y) \neq (0, 0)$ we have one of the alternatives:

$$\left.\begin{array}{ll} P(x, y): & x^2 + \beta xy + \gamma y^2 \geqslant 1, \\ N(x, y): & x^2 + \beta xy + \gamma y^2 \leqslant -1. \end{array}\right\} \tag{25}$$

If $P(1, -1)$ then $\gamma \geqslant \beta$, a contradiction to $2 \leqslant \beta \leqslant 3$ and $\beta^2 - 4\gamma > 0$. Hence $N(1, -1)$, that is,

$$-\beta + \gamma \leqslant -2.$$

If $P(0, 1)$, that is, $\gamma \geqslant 1$, this gives $\beta \geqslant 3$. Hence $\beta = 3, \gamma = 1$, since $2 \leqslant \beta \leqslant 3$, and so $f(x, y) = x^2 + 3xy + y^2$, the first Markoff form. Otherwise $N(0, 1)$, that is,

$$\gamma \leqslant -1.$$

Consider $f(-5, 2)$. If $P(-5, 2)$, then $25 - 10\beta + 4\gamma \geqslant 1$, and so, using also $N(0, 1)$ we have

$$10\beta \leqslant 24 + 4\gamma \leqslant 20.$$

Then $\beta = 2, \gamma = -1$ since $2 \leqslant \beta \leqslant 3$, which gives $x^2 + 2xy - y^2$, the second Markoff form. Otherwise, we have

$$\text{both} \quad N(0, 1) \quad \text{and} \quad N(-5, 2). \tag{26}$$

The proof now follows by induction. Let $(m, k: m_1, k_1; m_2, k_2)$ be an ordered Markoff set and suppose that we have

$$\text{both} \quad N(k_1, m_1) \quad \text{and} \quad N(k_2 - 3m_2, m_2). \tag{27}$$

Thus (26) is (27) for $(5, 2: 1, 0; 2, 1)$. We now consider the various possibilities for $f(k, m)$ and $f(k - 3m, m)$. First, if

$$\text{both} \quad P(k, m) \quad \text{and} \quad P(k - 3m, m),$$

we have $f(x, y) = F_m(x, y)$ by Lemma 8, Corollary. Otherwise we have at least one of

$$\text{both} \quad N(k, m) \quad \text{and} \quad N(k_2 - 3m_2, m_2), \tag{28}$$

or $\qquad \text{both} \quad N(k_1, m_1) \quad \text{and} \quad N(k - 3m, m). \tag{29}$

But (28) and (29) are just (27) for

$$(m_1', k_1': m, k; m_2, k_2) \quad \text{and} \quad (m_2', k_2': m_1, k_1; m, k),$$

respectively; where (m_1', m, m_2), (m_2', m_1, m) are just the two entries immediately below (m, m_1, m_2) in Figure 1. Hence, if

$f(x, y)$ were not a Markoff form it must satisfy (27) for an infinite sequence of Markoff sets

$$M^{(j)} = (m^{(j)}, k^{(j)}: m_1^{(j)}, k_1^{(j)}; m_2^{(j)}, k_2^{(j)}), \tag{30}$$

where $m^{(1)} < m^{(2)} < \dots$. But (27) implies $\beta^2 - 4\gamma \geqslant 9 - 4m^{-2}$ by Lemma 11, so

$$\beta^2 - 4\gamma \geqslant \lim_{j \to \infty} (9 - 4(m^{(j)})^{-2}) = 9,$$

contrary to hypothesis. This proves the lemma.

COROLLARY. *Let $m > 2$, $\tilde{m} > 2$ and suppose that $(\tilde{m}, \tilde{m}_1, \tilde{m}_2)$ is on the unique path in Figure 1 from $(1, 1, 1)$ down to (m, m_1, m_2). Then F_m satisfies (27) with $\tilde{M} = (\tilde{m}, \tilde{k}: \tilde{m}_1, \tilde{k}_1; \tilde{m}_2, \tilde{k}_2)$ substituted for $M = (m, k: m_1, k_1; m_2, k_2)$.*

Proof. $f(x, y) = F_m(x, y)$ satisfies the hypotheses of the lemma by (8), (9) and Lemma 10. Hence it must be thrown up by the preceding proof. The only way for this to happen is for the sequence (30) obtained to terminate with an $M^{(J)} = M$; so $\tilde{M} = M^{(j)}$ for some $j \leqslant J$.

LEMMA 14. *There are non-enumerably many forms*

$$f(x, y) = x^2 + \beta xy + \gamma y^2$$

with $2 \leqslant \beta \leqslant 3$ and $\beta^2 - 4\gamma = 9$ such that $|f(x, y)| \geqslant 1$ for all integers $(x, y) \neq (0, 0)$.

Proof. Let $\mathfrak{M}$ be any infinite sequence of $M^{(j)}$ ($j = 1, 2, \dots$) as in (30), where $(m^{(j)}, m_1^{(j)}, m_2^{(j)})$ is $(1, 1, 1)$, $(2, 1, 1)$, $(5, 1, 2)$ for $j = 1, 2, 3$ respectively, and $(m^{(j+1)}, m_1^{(j+1)}, m_2^{(j+1)})$ for $j \geqslant 3$ is one of the two solutions immediately below $(m^{(j)}, m_1^{(j)}, m_2^{(j)})$ in Figure 1. There are clearly non-enumerably many sequences $\mathfrak{M}$ and we shall show that to each there corresponds a distinct pair β, γ with the properties required. Put

$$F^{(j)} = F_{m^{(j)}} = x^2 + \beta^{(j)} xy + \gamma^{(j)} y^2.$$

Then

$$(\beta^{(j)})^2 - 4\alpha^{(j)} \gamma^{(j)} = 9 - 4(m^{(j)})^{-2} \to 9.$$

By the 'Compactness Lemma' 2 there are β, γ and a subsequence $j_1 < j_2 < \dots$ such that†

$$\beta^{(j_i)} \to \beta, \quad \gamma^{(j_i)} \to \gamma \quad \text{and} \quad \beta^2 - 4\gamma = 9.$$

---

† It is not difficult to see that in fact $\beta^{(j)}$, $\gamma^{(j)}$ tend to limits.

Put $f(x, y) = x^2 + \beta xy + \gamma y^2$. Then, by Lemma 10,

$$|f(x, y)| = \lim_{t \to \infty} |F^{(j_t)}(x, y)| \geqslant 1$$

for all integers $(x, y) \neq (0, 0)$. By Lemma 13, Corollary, $F^{(j)}(x, y)$ satisfies (27) for all sets $M^{(i)}$ $(3 \leqslant i \leqslant j)$. Hence in the limit, remembering the definition (25) of $N(x, y)$, we see that $f(x, y)$ satisfies (27) for all $M^{(i)}$ $(3 \leqslant i < \infty)$. But now two distinct sequences $\mathfrak{M}$, $\overline{\mathfrak{M}}$ (say) must have $M^{(j)}$, $\overline{M}^{(j)}$ which coincide for all j up to some J but differ for $j = J + 1$. Then one of the corresponding forms f, $\bar{f}$, say f, must satisfy (28) and the other, $\bar{f}$, must satisfy (29) for $M^{(J)}$. But (28) and (29) are incompatible by Lemma 12 since $\beta^2 - 4\gamma = 9$. Hence $f \neq \bar{f}$: that is to every sequence $\mathfrak{M}$ there corresponds a distinct form f.

5. The Markoff chain for forms.

THEOREM II. *Suppose that*

$$f(x, y) = \alpha x^2 + \beta xy + \gamma y^2, \quad \delta = \beta^2 - 4\alpha\gamma > 0,$$

and put

$$\mu = \inf |f(x, y)| \quad (x, y \ \textit{integers not both } 0).$$

A. *If* $\qquad\qquad \mu > \tfrac{1}{3}\delta^{\frac{1}{2}},$ \hfill (1)

then f is equivalent to a multiple of a Markoff form.

B. *Conversely* (1) *holds for all forms equivalent to multiples of Markoff forms.*

C. *There are non-enumerably many forms, none of which is equivalent to a multiple of any other, such that $\mu = \tfrac{1}{3}\delta^{\frac{1}{2}}$.*

Proof. By considering $\mu^{-1}f$ instead of f we may suppose that $\mu = 1$. Part B is merely a restatement of Lemma 10. Part C follows at once from Lemma 14 and the remark that there are only an enumerable set of forms $f'(x, y) = f(ax + by, cx + dy)$ with integers a, b, c, d equivalent to any given form f. Hence we may suppose that

$$0 < \delta < 9, \quad \mu = 1,$$

and it remains to prove A.

By hypothesis, given $\epsilon > 0$ there are integers a, c, which may be supposed coprime, such that $1 = \mu \leqslant |f(a,c)| < 1 + \epsilon$. Hence by Lemma 1, Corollary, there is a form

$$f'(x,y) = \alpha' x^2 + \beta' xy + \gamma' y^2$$

equivalent to $\pm f(x, y)$, and such that

$$1 \leqslant \alpha' < 1 + \epsilon, \quad 2\alpha' \leqslant \beta' \leqslant 3\alpha'.$$

If $\alpha' = 1$, then $f'(x, y)$ is a Markoff form by Lemma 13; and the conclusion is true, since $\pm F_m(x, y)$ are equivalent by Lemma 9. Otherwise we may find an infinite sequence of forms

$$f_n(x,y) = \alpha_n x^2 + \beta_n xy + \gamma_n y^2,$$
$$\alpha_n \to 1, \quad 2\alpha_n \leqslant \beta_n \leqslant 3\alpha_n, \quad \beta_n^2 - 4\alpha_n \gamma_n = \delta, \qquad (2)$$

each equivalent to $\pm f(x,y)$ so that, in particular,

$$|f_n(x,y)| \geqslant 1, \quad (x,y) \text{ integral} \neq (0,0).$$

By the 'Compactness Lemma' 2 we may suppose, by taking a subsequence of the original subsequence if necessary, that β_n, γ_n tend to limits, say

$$\alpha_n \to 1 = \alpha_0, \quad \beta_n \to \beta_0, \quad \gamma_n \to \gamma_0, \qquad (3)$$

with $\qquad\qquad 2 \leqslant \beta_0 \leqslant 3, \quad \beta_0^2 - 4\gamma_0 = \delta.$

Write $\qquad\qquad f_0(x,y) = x^2 + \beta_0 xy + \gamma_0 y^2.$

Let θ_0, ϕ_0 be the roots of $f_0(x, 1)$ and† θ_n, ϕ_n those of $f_n(x, 1)$; so

$$\theta_n \to \theta_0, \quad \phi_n \to \phi_0, \qquad (4)$$

with an appropriate choice for θ_n between θ_n, ϕ_n. Further,

$$|f_0(x,y)| = \lim |f_n(x,y)| \geqslant 1 \quad (n \to \infty),$$

for all integers $(x, y) \neq (0, 0)$. Hence, by Lemma 13, $f_0(x, y)$ is a $F_m(x, y)$. But $F_m(x, y)$ takes both values ± 1 by Lemma 8 and so the 'Isolation Theorem' I applies to it. Let $\mu' < \mu = 1$, $\epsilon_0 > 0$ be the corresponding constants given by Theorem I for $f_0(x, y) = F_m$. Then
$$|\alpha_n - \alpha_0| < \epsilon_0, \quad |\theta_n - \theta_0| < \epsilon_0, \quad |\phi_n - \phi_0| < \epsilon_0,$$

† The notation θ_n, ϕ_n is not to be confused with that of the previous chapter.

for some sufficiently large n by (3) and (4). Since, in fact,

$$|f_n(x, y)| \geqslant 1 > \mu'$$

for all integers $(x, y) \neq (0, 0)$, it follows that $f_n(x, y)$ is $\lambda F_m(x, y)$ for some constant† λ. But $\pm f$ is equivalent to f_n; and so A is true.

6. The Markoff chain for approximations.

THEOREM III. *Let θ be irrational and put*

$$\nu = \liminf q \, \| q\theta \|. \tag{1}$$

A. *If $\nu > \frac{1}{3}$ then θ is equivalent to a root of $F_m(x, 1) = 0$, where F_m is a Markoff form.*

B. *Conversely if θ is equivalent to a root of $F_m(x, 1) = 0$ then*

$$\nu = (9 - 4m^{-2})^{-\frac{1}{2}} > \tfrac{1}{3}, \tag{2}$$

and there are infinitely many solutions of $q \, \| q\theta \| < \nu$. The two roots of $F_m(x, 1) = 0$ are equivalent one to the other.

C. *There are non-enumerably many inequivalent θ such that $\nu = \frac{1}{3}$.*

Proof of A. We write

$$f(x, y) = x(\theta x - y). \tag{3}$$

By hypothesis, given any $\epsilon > 0$, however small, there is an $X_0 = X_0(\epsilon)$ such that

$$|f(x, y)| > \nu - \epsilon \quad \text{whenever} \quad |x| > X_0 \tag{4}$$

and x, y are integers. Since θ is irrational, a moment's thought shows that this is equivalent to the statement that there is a $Y_0 = Y_0(\epsilon) > 0$ such that

$$|f(x, y)| > \nu - \epsilon \quad \text{whenever} \quad |\theta x - y| < Y_0(\epsilon) \tag{5}$$

and $(x, y) \neq (0, 0)$ are integers.

Further, by hypothesis, there is a sequence of pairs of integers a_n, b_n, which may be supposed coprime, such that

$$|f(a_n, b_n)| \to \nu, \quad a_n \to \infty, \quad |\theta a_n - b_n| \to 0. \tag{6}$$

† It is easy to see that $\lambda = 1$.

By Lemma 1, Corollary, there are substitutions

$$x_n = a_n x + c_n y, \quad y_n = b_n x + d_n y; \tag{7}$$

$$a_n, b_n, c_n, d_n \text{ integers}, \quad a_n d_n - b_n c_n = \pm 1, \tag{8}$$

such that

$$\pm f(x_n, y_n) = f_n(x, y) = \alpha_n x^2 + \beta_n xy + \gamma_n y^2 \tag{9}$$

satisfies

$$\left. \begin{aligned} \alpha_n &= |f(a_n, b_n)| \to \nu, \\ 2\alpha_n \leqslant \beta_n \leqslant 3\alpha_n, \quad \beta_n^2 - 4\alpha_n \gamma_n &= 1. \end{aligned} \right\} \tag{10}$$

By the 'Compactness Lemma' 2 we may suppose, by taking a subsequence, that

$$\left. \begin{aligned} \alpha_n \to \alpha_0 = \nu, \quad \beta_n \to \beta_0, \quad \gamma_n \to \gamma_0; \\ 2\alpha_0 \leqslant \beta_0 \leqslant 3\alpha_0, \quad \beta_0^2 - 4\alpha_0 \gamma_0 = 1, \end{aligned} \right\} \tag{11}$$

for some β_0, γ_0. Write

$$\begin{aligned} f_0(x, y) &= \alpha_0 x^2 + \beta_0 xy + \gamma_0 y^2 \\ &= \nu(x - \theta_0 y)(x - \phi_0 y). \end{aligned} \tag{12}$$

Write

$$\theta_n = \frac{-\theta c_n + d_n}{\theta a_n - b_n}, \quad \phi_n = \frac{-c_n}{a_n}. \tag{13}$$

Then, by (7),

$$x_n = a_n(x - \phi_n y), \tag{14}$$

$$\theta x_n - y_n = (\theta a_n - b_n)(x - \theta_n y). \tag{15}$$

Hence

$$f_n(x, y) = \alpha_n(x - \theta_n y)(x - \phi_n y). \tag{16}$$

By interchanging θ_0, ϕ_0 and taking only a subsequence of the f_n if need be, we may suppose that

$$\theta_n \to \theta_0, \quad \phi_n \to \phi_0. \tag{17}$$

Let x, y be fixed integers and x_n, y_n be defined by (7). Then

$$\begin{aligned} \lim |\theta x_n - y_n| &= \lim |x - \theta_n y| \, |\theta a_n - b_n| \\ &= 0 \quad (n \to \infty), \end{aligned} \tag{18}$$

by (6), (15), since $|x - \theta_n y|$ is bounded. Hence

$$\begin{aligned} |f_0(x, y)| &= \lim |f_n(x, y)| \quad (n \to \infty) \\ &= \lim |f(x_n, y_n)| \\ &\geqslant \nu \end{aligned} \tag{19}$$

by (5). Thus, and by (11), $\nu^{-1}f_0(x,y)$, which has discriminant $\nu^{-2} < 9$ by hypothesis, satisfies the conditions of Lemma 13; so

$$f_0(x,y) = \nu F_m(x,y) \tag{20}$$

for some Markoff form F_m.

If $\theta_n = \theta_0$ for any n it follows that θ is equivalent to θ_0; and so A is true. We may thus suppose that

$$\theta_n \neq \theta_0 \quad \text{(all } n \geqslant 1), \tag{21}$$

and will deduce a contradiction. Clearly Theorem I, Corollary, applies with $\mu = 1$ to $F_m(x,y)$ and $\nu^{-1}f_n(x,y)$ by (17), (18), (20), (21). Hence there is a $\mu' < 1$ such that for all sufficiently large n there are integers $(\bar{x}_n, \bar{y}_n)$ with

$$|f_n(\bar{x}_n, \bar{y}_n)| < \mu'\nu, \quad |\bar{x}_n - \theta_n\bar{y}_n| \leqslant 1. \tag{22}$$

Put
$$\bar{x}^{(n)} = a_n\bar{x}_n + c_n\bar{y}_n, \quad \bar{y}^{(n)} = b_n\bar{x}_n + d_n\bar{y}_n.$$

Then, by (9),

$$|f(\bar{x}^{(n)}, \bar{y}^{(n)})| = |f_n(\bar{x}_n, \bar{y}_n)| < \mu'\nu < \nu,$$

and, by (6), (15), (22),

$$|\theta\bar{x}^{(n)} - \bar{y}^{(n)}| = |\theta a_n - b_n| \, |\bar{x}_n - \theta_n\bar{y}_n|$$
$$\leqslant |\theta a_n - b_n|$$
$$\to 0 \quad (n \to \infty).$$

This is a contradiction to (5).

Proof of B. Follows at once from Lemmas 4, 9 (with its Corollary), 10 since then $F_m(x,y)$ has $\mu = 1$, $\delta = 9 - 4m^{-2}$.

Proof of C. By Lemmas 4, 14 there are non-enumerably many θ with $\nu \geqslant \frac{1}{3}$. By A there are only enumerably many θ with $\nu > \frac{1}{3}$ and it is clear that the set of numbers equivalent to any number is enumerable.

NOTES

§ 1. Markoff's original proof used continued fractions (MARKOFF (1879) or see DICKSON (1930)). The present proof goes back to REMAK (1924) and FROBENIUS (1913). For another point of view (but no proof) see COHN (1955).

Not much is known about possible values below $\frac{1}{3}$ of $\delta^{-\frac{1}{2}}(f)\,\mu(f)$ or, what is practically the same, of $\nu(\theta)$; see KOKSMA (1936), Kap III. For a simple proof that values between $12^{-\frac{1}{2}}$ and $13^{-\frac{1}{2}}$ cannot occur, see DAVIS (1950). Marshall Hall has shown that they take all values in an interval to the right of 0 but has not published the details: a result of this kind is a fairly immediate deduction from (2·15) of Chapter I and the results of M. HALL (1947), where such an application is adumbrated.

§ 2. The 'Isolation Theorem' and its application in this context are due to C. A. Rogers (unpublished). There is a curious anticipation in REMAK (1925). For an extension of the 'isolation' technique see CASSELS & SWINNERTON-DYER (1955).

§ 3. For a further discussion of the equation see FROBENIUS (1913). Frobenius's technique has recently been applied to other Diophantine equations by several workers.

CHAPTER III

INHOMOGENEOUS APPROXIMATION

1. Introduction. In the preceding two chapters we have
been concerned with homogeneous problems—that is, we have
tried to make the fractional part $\| q\theta \|$ of the homogeneous ex-
pression $q\theta$ small, or more generally we have tried to make
$\| q\theta_1 \|, ..., \| q\theta_n \|$ simultaneously small. In this chapter we shall
be concerned with the inhomogeneous form $q\theta - \alpha$, or with
more general simultaneous problems. There is at once a point
of difference. In the homogeneous problem the value $q = 0$
gives a trivial result which must be excluded and there is no
additional generality in allowing $q < 0$ since $\| (-q)\theta \| = \| q\theta \|$.
In the inhomogeneous case it is usually appropriate to allow
the integer variables to take all values positive, negative or
zero. The restriction to positive values leads to a different kind
of problem.

If θ is rational, say $\theta = m/n$, where $n > 0$, m are integers,
it is trivial that $\| q\theta - \alpha \| \geqslant n^{-1} \| n\alpha \|$ with equality for infinitely
many q. Another trivial case is when $\alpha = m\theta + n$ for some
integers m, n. Then $\| q\theta - \alpha \| = \| (q-m)\theta \|$ and so the problem
of the behaviour of $\| q\theta - \alpha \|$ is essentially a homogeneous one.
In § 2 we shall show that, apart from these two cases, there are
always infinitely many integers q such that $| q | \| q\theta - \alpha \| < \frac{1}{4}$;
and that for variable θ the $\frac{1}{4}$ in this statement cannot be replaced
by a smaller constant. This theorem is analogous to the homo-
geneous theorem about $q \| q\theta \| < 5^{-\frac{1}{2}}$. In § 3 we show that
there is no possible inhomogeneous analogue, however weak,
to the existence of a solution of $0 < q < Q$, $\| q\theta \| \leqslant Q^{-1}$ for
all $Q > 1$.

For simultaneous inhomogeneous approximation new con-
siderations arise. Suppose that we wish to find an integer q
such that simultaneously

$$\| q\theta_i - \alpha_i \| < \epsilon \quad (1 \leqslant i \leqslant n), \tag{1}$$

where θ_i, α_i and $\epsilon > 0$ are given. Let there be integers

$u_1, \ldots, u_n$ not all 0, such that $u_1\theta_1 + \ldots + u_n\theta_n$ is an integer. Then (1) implies that

$$\| u_1\alpha_1 + \ldots + u_n\alpha_n \|$$
$$= \| u_1(\alpha_1 - q\theta_1) + \ldots + u_n(\alpha_n - q\theta_n) \|$$
$$\leqslant | u_1 | \, \| \alpha_1 - q\theta_1 \| + \ldots + | u_n | \, \| \alpha_n - q\theta_n \|$$
$$< (| u_1 | + \ldots + | u_n |)\epsilon.$$

Thus if (1) is to be soluble for arbitrarily small $\epsilon > 0$ we must have $\| u_1\alpha_1 + \ldots + u_n\alpha_n \| = 0$, that is, $u_1\alpha_1 + \ldots + u_n\alpha_n$ must be an integer. This gives a set of necessary conditions that (1) be soluble for arbitrarily small $\epsilon > 0$. In § 5 we shall show, in a rather more general context, that this necessary condition is also sufficient. In Chapter V we shall discuss to what extent this result can be formulated quantitatively.

2. The 1-dimensional case. It is convenient first to prove a result of a rather general kind:

THEOREM I. (Minkowski.)   *Let* $L_j = L_j(x, y) = \lambda_j x + \mu_j y$ *for* $j = 1, 2$ *be a pair of linear forms and let* $\Delta = \lambda_1\mu_2 - \lambda_2\mu_1 \neq 0$. *Then:*

A. *For all numbers* ρ_1, ρ_2 *there are integers* x, y *such that*

$$| L_1 + \rho_1 | \, | L_2 + \rho_2 | \leqslant \tfrac{1}{4} | \Delta |. \tag{1}$$

B. *If, further,* μ_1/λ_1 *is irrational and* $\epsilon > 0$ *is arbitrarily small there are solutions of* (1) *with*

$$| L_1 + \rho_1 | < \epsilon. \tag{2}$$

Theorem I is an easy consequence of:

LEMMA 1.   *Let* $\theta, \phi, \psi, \omega$ *be four real numbers such that*

$$| \theta\omega - \phi\psi | \leqslant \tfrac{1}{2} | \Delta |, \quad | \psi\omega | \leqslant | \Delta |, \quad \psi > 0. \tag{3}$$

Then there is an integer u *such that*

$$| \theta + \psi u | \, | \phi + \omega u | \leqslant \tfrac{1}{4} | \Delta |, \tag{4}$$

and $$| \theta + \psi u | \leqslant \psi. \tag{5}$$

Proof. By writing $\theta + u_0\psi$, $\phi + u_0\omega$ for θ, ϕ respectively, where u_0 is a suitable integer, we may suppose that

$$-\psi \leqslant \theta < 0,$$

and, by writing $-\phi$, $-\omega$ for ϕ, ω if necessary, that

$$\phi \geqslant 0.$$

We show that $u = 0$ or 1 will now do.

Suppose first that $\phi + \omega \leqslant 0$. Then

$$
\begin{aligned}
16 \,|\, \theta\phi \,|\; |\, (\theta+\psi)(\phi+\omega) \,| \\
\leqslant (|\,\theta\,| + |\,\theta+\psi\,|)^2 (|\,\phi\,| + |\,\phi+\omega\,|)^2 \\
= \psi^2\omega^2 \leqslant |\,\Delta\,|^2.
\end{aligned}
\tag{6}
$$

If, however, $\phi + \omega \geqslant 0$, then

$$
\begin{aligned}
2(|\,\theta\phi\,|\;|\,(\theta+\psi)(\phi+\omega)\,|)^{\frac{1}{2}} &\leqslant |\,\phi\,|\;|\,\theta+\psi\,| + |\,\theta\,|\;|\,\phi+\omega\,| \\
&= |\,\phi(\theta+\psi) - \theta(\phi+\omega)\,| \\
&= |\,\phi\psi - \theta\omega\,| \leqslant \tfrac{1}{2}|\,\Delta\,|.
\end{aligned}
\tag{7}
$$

Hence in either case

$$\min(|\,\theta\phi\,|, \; |\,(\theta+\psi)(\phi+\omega)\,|) \leqslant \tfrac{1}{4}|\,\Delta\,|. \tag{8}$$

Since $\max(|\,\theta\,|, |\,\theta+\psi\,|) \leqslant \psi$, this proves the lemma.

Proof of Theorem I. We commence with B. By Minkowski's linear forms theorem (Appendix B, Theorem III) there are integers x_0, y_0, not both 0, such that

$$|\,\lambda_1 x_0 + \mu_1 y_0\,| < \epsilon, \quad |\,\lambda_2 x_0 + \mu_2 y_0\,| \leqslant \epsilon^{-1}|\,\Delta\,|. \tag{9}$$

Without loss of generality x_0, y_0 are coprime. Since μ_1/λ_1 is irrational we may suppose, by taking $-x_0$, $-y_0$ if necessary, that

$$0 < \lambda_1 x_0 + \mu_1 y_0 < \epsilon. \tag{9'}$$

Choose now integers x_1, y_1 such that $x_0 y_1 - x_1 y_0 = 1$. Put

$$x = x'x_0 + y'x_1, \quad y = x'y_0 + y'y_1,$$

so that x, y are integers if x', y' are and vice versa. Then

$$L_j = \lambda_j x + \mu_j y = \lambda_j' x' + \mu_j' y' \quad (j = 1, 2),$$

where $\qquad \lambda_1'\mu_2' - \lambda_2'\mu_1' = \lambda_1\mu_2 - \lambda_2\mu_1 = \Delta,$

and $\qquad 0 < \lambda_1' < \epsilon, \quad |\,\lambda_2'\,| \leqslant \epsilon^{-1}|\,\Delta\,|,$

by (9) and (9').

Now let y' be an integer such that

$$|\rho_1\lambda_2' - \rho_2\lambda_1' - \Delta y'| \leqslant \tfrac{1}{2}|\Delta|.$$

We may now apply Lemma 1 with

$$\theta = \mu_1'y' + \rho_1, \quad \phi = \mu_2'y' + \rho_2,$$
$$\psi = \lambda_1', \quad \omega = \lambda_2',$$

since $\quad |\theta\omega - \psi\phi| = |\rho_1\lambda_2' - \rho_2\lambda_1' - \Delta y'| \leqslant \tfrac{1}{2}|\Delta|$

and $\quad\quad\quad |\psi\omega| = |\lambda_1'\lambda_2'| < |\Delta|.$

Hence there exists an integer $x' = u$ such that

$$L_j = \lambda_j'x' + \mu_j'y' \quad (j = 1, 2),$$

satisfy $\quad\quad |L_1 + \rho_1||L_2 + \rho_2| \leqslant \tfrac{1}{4}|\Delta|,$

$$|L_1 + \rho_1| \leqslant \lambda_1' < \epsilon,$$

as required.

To prove A we note that there are certainly integers x_0, y_0 not both 0, such that

$$|\lambda_j x_0 + \mu_j y_0| \leqslant |\Delta|^{\frac{1}{2}}, \quad (j = 1, 2).$$

If $\lambda_1 x_0 + \mu_1 y_0 \neq 0$ we proceed as above, but if $\lambda_1 x_0 + \mu_1 y_0 = 0$ then $\lambda_2 x_0 + \mu_2 y_0 \neq 0$ and the roles of L_1 and L_2 are to be interchanged.

COROLLARY. *The constant $\tfrac{1}{4}$ cannot be replaced by a smaller one.*
Proof.
$$|x + \tfrac{1}{2}||y + \tfrac{1}{2}| \geqslant \tfrac{1}{4}$$

for all integers x, y.

It could be shown by examining the proof more closely that this is essentially the only case when equality is required in (1). We prove now instead

THEOREM II A. (Minkowski.) *If θ is irrational and α is not of the form $\alpha = m\theta + n$ for integers m, n then there are infinitely many integers q such that*
$$|q|\,\|q\theta - \alpha\| < \tfrac{1}{4}.$$

B. *For any given $\epsilon > 0$ there is an irrational θ and an α not of the form $m\theta + n$ such that $|q|\,\|q\theta - \alpha\| > \tfrac{1}{4} - \epsilon$ for all $q \neq 0$ and $\liminf |q|\,\|q\theta - \alpha\| = \tfrac{1}{4}$ as $|q| \to \infty$.*

Proof of A. By Theorem I B with

$$L_1 + \rho_1 = \theta x - y - \alpha, \quad L_2 + \rho_2 = x, \quad |\Delta| = 1,$$

there are integers $x = q$, $y = p$ such that

$$|q| \, |q\theta - p - \alpha| \leqslant \tfrac{1}{4}, \quad |q\theta - p - \alpha| < \epsilon.$$

Since $\alpha \neq q\theta - p$ for all integers p, q we must get infinitely many pairs of integers p, q by letting $\epsilon \to 0$. Finally $\tfrac{1}{4}$ can occur at most once since

$$q\theta - p - \alpha = \pm \tfrac{1}{4} q^{-1}, \quad q'\theta - p' - \alpha = \pm \tfrac{1}{4} q'^{-1}, \quad q \neq q',$$

(independent signs) implies that $(q - q')\theta$, and so θ, is rational contrary to hypothesis.

Proof of B. Write

$$\theta = [a_1, a_2, \ldots]$$

in the continued fraction notation of Chapter I, where the a_n will be subjected to conditions in the course of the proof. We have

$$\left| \frac{q_{n+1}\theta - p_{n+1}}{q_n\theta - p_n} \right| = \theta_{n+1} = [a_{n+1}, a_{n+2}, \ldots] < a_{n+1}^{-1} \tag{10}$$

and, for $n \geqslant 1$,

$$\frac{q_n}{q_{n+1}} = \phi_n = [a_n, a_{n-1}, \ldots, a_1] \leqslant a_n^{-1}. \tag{11}$$

By (2·15), (2·16) of Chapter I we have†

$$|q_n(q_n\theta - p_n)| = (a_n + \phi_{n-1} + \theta_{n+1})^{-1}$$
$$= (a_n + O(1))^{-1}, \tag{12}$$

and

$$|q_{n+1}(q_n\theta - p_n)| = (1 + \theta_{n+1}\phi_n)^{-1}$$
$$= 1 + O(a_n^{-1}a_{n+1}^{-1}). \tag{13}$$

Now put

$$\alpha = \tfrac{1}{2}(1 - \theta).$$

It is clearly enough to consider only integers $q \neq 0$, p such that

$$1 \geqslant 4|q| \, |q\theta - p - \alpha| = |2q| \, |(2q+1)\theta - (2p+1)|. \tag{14}$$

† $f = O(g)$ means that $|fg^{-1}|$ is less than an absolute constant, where $g > 0$, f may depend on a number of variables: and, for example, $f = h + O(g)$ means $f - h = O(g)$.

CDA

If $\left|\,2q+1\,\right| \leqslant a_1^{\frac{1}{2}}$ then $\left|\,(2q+1)\,\theta\,\right| < a_1^{-\frac{1}{2}}$ and the right-hand side of (14) is

$$> 2(1 - a_1^{-\frac{1}{2}}) \geqslant 1$$

when $a_1 \geqslant 4$, which we now assume. There is thus an integer $n \geqslant 1$ such that

$$a_n^{\frac{1}{2}} q_n \leqslant \left|\,2q+1\,\right| \leqslant a_{n+1}^{\frac{1}{2}} q_{n+1}. \tag{15}$$

Hence, by (12), (14), and the trivial $\left|\,(2q+1)/2q\,\right| < 2$,

$$\frac{\left|\,(2q+1)\,\theta - (2p+1)\,\right|}{\left|\,q_n \theta - p_n\,\right|} \leqslant \frac{\left|\,2q+1\,\right|}{\left|\,2q\,\right|} \cdot \frac{q_n}{\left|\,2q+1\,\right|} \cdot \frac{1}{\left|\,q_n(q_n\theta - p_n)\,\right|}$$
$$= O(a_n^{\frac{1}{2}}). \tag{16}$$

Since $\left|\,p_{n+1}q_n - q_{n+1}p_n\,\right| = 1$ there are integers u, v such that

$$2p+1 = up_n + vp_{n+1}, \quad 2q+1 = uq_n + vq_{n+1}. \tag{17}$$

Indeed, by (15), (16),

$$\left|\,u\,\right| = \left|\,(2q+1)\,(q_{n+1}\theta - p_{n+1}) - q_{n+1}((2q+1)\,\theta - (2p+1))\,\right|$$
$$= O(a_{n+1}^{\frac{1}{2}} q_{n+1} \left|\,q_{n+1}\theta - p_{n+1}\,\right|) + O(a_n^{\frac{1}{2}} q_{n+1} \left|\,q_n\theta - p_n\,\right|)$$
$$= O(a_n^{\frac{1}{2}}),$$

on using (12), (13). Thus

$$\frac{2q+1}{q_{n+1}} = v + u\frac{q_n}{q_{n+1}} = v + O(a_n^{-\frac{1}{2}}). \tag{18}$$

Hence and by (15) we have $v = O(a_{n+1}^{\frac{1}{2}})$; and so

$$\frac{(2q+1)\,\theta - (2p+1)}{q_n\theta - p_n} = u + v\frac{q_{n+1}\theta - p_{n+1}}{q_n\theta - p_n} = u + O(a_{n+1}^{-\frac{1}{2}}). \tag{19}$$

We now suppose that all a_n are even. Since

$$q_{n+1} = a_n q_n + q_{n-1}, \quad p_{n+1} = a_n p_n + p_{n-1}$$

and $\qquad\qquad p_0 = q_1 = 1, \quad p_1 = q_0 = 0,$

either p_n, q_{n+1} are odd and q_n, p_{n+1} are even or vice versa. In either case u, v are odd by (17), so $uv \neq 0$. Thus, by (18), (19),

$$\frac{\left|\,2q+1\,\right| \left|\,(2q+1)\,\theta - (2p+1)\,\right|}{q_{n+1} \left|\,q_n\theta - p_n\,\right|} \geqslant 1 - O(a_n^{-\frac{1}{2}}) - O(a_{n+1}^{-\frac{1}{2}}).$$

But $|\,2q/(2q+1)\,| \geqslant 1-O(q_n^{-1}a_n^{-\frac{1}{2}})$ by (15); and so, by (13),

$$4\,|\,q(q\theta-p-\alpha)\,| = |\,2q\,|\,|\,(2q+1)\,\theta-(2p+1)\,|$$
$$> 1-O(a_n^{-\frac{1}{2}})-O(a_{n+1}^{-\frac{1}{2}}).$$

This is $>1-4\epsilon$ if $\min a_n$ is larger than a constant depending only on ϵ. Clearly $n\to\infty$ as $|\,q\,|\to\infty$, and so

$$\lim\inf|\,q(q\theta-p-\alpha)\,| \geqslant \tfrac{1}{4}$$

if, further, $a_n\to\infty$. The constructed θ, α have thus the properties required.

3. A negative result.

THEOREM III. *Let $\phi(q)$ be any positive function of the integer variable q such that*

$$\phi(q)\to 0 \quad (q\to\infty). \tag{1}$$

Then there is an α and an irrational θ such that the pair of inequalities

$$|\,q\,| \leqslant Q, \quad \|\,q\theta-\alpha\,\| < \phi(Q) \tag{2}$$

is insoluble for infinitely many values of Q.

Note. $\phi(q)$ may tend to 0 arbitrarily slowly. This is then a contrast with the fact that $0<q<Q$, $\|\,q\theta\,\| \leqslant Q^{-1}$ is always soluble. In our example α is rational but it is not difficult to modify the construction so as to give an irrational α.

Proof. We put $\alpha=\tfrac{1}{2}$ and choose θ as the limit of a sequence of rational numbers u_n/v_n $(n=1,2,\ldots)$ where u_n, v_n are integers and v_n is odd. Hence

$$\left|\,q\frac{u_n}{v_n}-\frac{1}{2}\,\right| \geqslant \frac{1}{2v_n}, \tag{3}$$

for all integers q. We shall also define integers Q_n for $n\geqslant 2$. Put $u_1/v_1=\tfrac{1}{3}$. If u_n/v_n, Q_n have been defined for $n\leqslant N$ we define Q_{N+1} to be any integer such that

$$\begin{aligned}\phi(Q_{N+1}) &< (4v_N)^{-1} \quad (N\geqslant 1),\\ Q_{N+1} &> 2Q_N \qquad\quad (N\geqslant 2),\end{aligned} \tag{4}$$

which is possible by (1). We then take u_{N+1}, v_{N+1} to be any integers such that v_{N+1} is odd and

$$\left|\,\frac{u_{N+1}}{v_{N+1}}-\frac{u_N}{v_N}\,\right| < \frac{1}{8v_NQ_{N+1}}, \quad v_{N+1}>2v_N.$$

Then $\qquad\qquad \theta = \lim u_n/v_n \quad (n \to \infty)$

exists, and indeed

$$\left| \theta - \frac{u_n}{v_n} \right| < \frac{1}{8v_n Q_{n+1}} + \frac{1}{8v_{n+1} Q_{n+2}} + \dots$$

$$< \frac{1}{8v_n Q_{n+1}} (1 + \tfrac{1}{4} + \tfrac{1}{16} + \dots)$$

$$< \frac{1}{4v_n Q_{n+1}}.$$

Hence if $|q| \leqslant Q_{n+1}$ we have

$$\| q\theta - \tfrac{1}{2} \| \geqslant \left| q\frac{u_n}{v_n} - \frac{1}{2} \right| - q\left| \theta - \frac{u_n}{v_n} \right| \geqslant \frac{1}{2v_n} - \frac{1}{4v_n} > \phi(Q_{n+1}),$$

by (3) and (4). This proves the theorem for the given θ, α and for the infinitely many values $Q_2, Q_3, \dots$.

4. Linear independence over the rationals.

We shall say that a set of numbers $\mu_1, \dots, \mu_l$ is LINEARLY INDEPENDENT (OVER THE RATIONALS) if the only set of rational numbers $v_1, \dots, v_l$ for which $v_1 \mu_1 + \dots + v_l \mu_l = 0$ is $v_1 = \dots = v_l = 0$. Further, a number λ is LINEARLY DEPENDENT ON $\mu_1, \dots, \mu_l$ (OVER THE RATIONALS) if $\lambda = v_1 \mu_1 + \dots + v_l \mu_l$ for rational $v_1, \dots, v_l$. For the proof of Theorem IV we require

LEMMA 2. *Let* $\lambda_1, \dots, \lambda_n$ *be real numbers not all* 0. *There is a linearly independent set of numbers* $\mu_1, \dots, \mu_l$ $(l \leqslant n)$ *such that each* λ_j *is linearly dependent on* $\mu_1, \dots, \mu_l$.

Proof. If $\lambda_1, \dots, \lambda_n$ are linearly independent put $l = n$, $\mu_j = \lambda_j$. Otherwise there is an equation

$$v_1 \lambda_1 + \dots + v_n \lambda_n = 0, \qquad\qquad (1)$$

with rational $v_1, \dots, v_n$ not all 0. Without loss of generality $v_n \neq 0$, so λ_n is linearly dependent on $\lambda_1, \dots, \lambda_{n-1}$. If $\lambda_1, \dots, \lambda_{n-1}$ are linearly independent put $l = n-1$, $\mu_j = \lambda_j$ $(j \leqslant n-1)$. Otherwise there is a linear relation $v_1 \lambda_1 + \dots + v_{n-1} \lambda_{n-1} = 0$ and $v_{n-1} \neq 0$ without loss of generality. By repeating the previous argument we see that λ_n, λ_{n-1} are linearly dependent on

$\lambda_1, ..., \lambda_{n-2}$. Ultimately we obtain a linearly independent subset $\mu_j = \lambda_j$ $(1 \leqslant j \leqslant l)$, with a suitable ordering of the λ_j, on which $\lambda_1, ..., \lambda_n$ are linearly dependent.

COROLLARY. *If* $\lambda_1 \neq 0$ *then we can even choose* $\mu_1 = \lambda_1$.

Proof. For if $v_1 \neq 0$ in (1) then $v_j \neq 0$ for some $j \neq 1$, and we can achieve $v_n \neq 0$ by rearranging $\lambda_2, ..., \lambda_n$ only. Similarly at the later stages in the argument.

5. Simultaneous approximation (Kronecker's theorem).

THEOREM IV. (Kronecker.)† *Let*

$$L_j(\mathbf{x}) = L_j(x_1, ..., x_m) \quad (1 \leqslant j \leqslant n),$$

be n *homogeneous linear forms in any number* m *of variables* x_i. *Then each of the following two statements about a real vector* $\boldsymbol{\alpha} = (\alpha_1, ..., \alpha_n)$ *implies the other.*

A. *For each* $\epsilon > 0$ *there is an integral vector* $\mathbf{a} = (a_1, ..., a_m)$ *such that simultaneously*

$$\| L_j(\mathbf{a}) - \alpha_j \| < \epsilon \quad (1 \leqslant j \leqslant n). \tag{1}$$

B. *If* $\mathbf{u} = (u_1, ..., u_n)$ *is any integral vector such that*

$$u_1 L_1(\mathbf{x}) + ... + u_n L_n(\mathbf{x})$$

has integer coefficients, considered as a form in the indeterminates x_i, *then*

$$u_1 \alpha_1 + ... + u_n \alpha_n = \text{integer}. \tag{2}$$

That A implies B is trivial: it was proved already in §1 for a special case and the general proof is quite similar. It remains only to show that B implies A.

We denote by Λ the set of all $\mathbf{z} = (z_1, ..., z_n)$ which can be written in the form

$$z_j = L_j(\mathbf{x}) - y_j, \tag{3}$$

where $\mathbf{x} = (x_1, ..., x_m)$, $\mathbf{y} = (y_1, ..., y_n)$ are integers. Clearly if $\mathbf{z}^{(1)}$, $\mathbf{z}^{(2)} \in \Lambda$ then $a\mathbf{z}^{(1)} + b\mathbf{z}^{(2)} \in \Lambda$ for all integers a, b. All points $\mathbf{z}$ with integer $z_1, ..., z_n$ are in Λ. It is clear that if $\mathbf{u} = (u_1, ..., u_n)$ is integral the necessary and sufficient condition that

$$u_1 L_1(\mathbf{x}) + ... + u_n L_n(\mathbf{x})$$

† We use vector notation, see p. ix. There is another proof of Theorem IV at the end of Chapter V §8.

be a form in $\mathbf{x}$ with integral coefficients is that $\mathbf{u}\mathbf{z} =$ integer for all $\mathbf{z} \in \Lambda$. If $\mathbf{u}\mathbf{z} =$ integer for some real $\mathbf{u}$ and all $\mathbf{z} \in \Lambda$, then $u_1, \ldots, u_n$ are integers since each vector $(0, \ldots, 0, 1, 0, \ldots, 0) \in \Lambda$.

We may thus reformulate what is to be proved as follows:—Suppose that $\mathbf{u}\boldsymbol{\alpha} = u_1\alpha_1 + \ldots + u_n\alpha_n =$ integer for every real $\mathbf{u}$ such that $\mathbf{u}\mathbf{z} =$ integer for all $\mathbf{z} \in \Lambda$; then for any $\epsilon > 0$ there is a $\mathbf{z}^{(\epsilon)} \in \Lambda$ such that

$$|z_j^{(\epsilon)} - \alpha_j| < \epsilon \quad (1 \leqslant j \leqslant n).$$

LEMMA 3. *There is a set of $s \leqslant n$ integral vectors $\mathbf{u}^{(t)}$ $(1 \leqslant t \leqslant s)$ such that*

(i) *The necessary and sufficient condition that the real vector $\mathbf{u}$ satisfies $\mathbf{u}\mathbf{z} =$ integer for all $\mathbf{z} \in \Lambda$, is that $\mathbf{u} = v_1\mathbf{u}^{(1)} + \ldots + v_s\mathbf{u}^{(s)}$ for some integers $v_1, \ldots, v_s$.*

(ii) *After an appropriate permutation of the forms $L_j(\mathbf{x})$ if need be, the $\mathbf{u}^{(t)}$ take the form*

$$\mathbf{u}^{(t)} = (0, \ldots, 0, u_{t,t}, u_{t,t+1}, \ldots, u_{tn}) \quad u_{t,t} \neq 0.$$

Note. The infinitely many conditions $\mathbf{u}\boldsymbol{\alpha} =$ integer thus reduce to the finitely many $\mathbf{u}^{(t)}\boldsymbol{\alpha} =$ integer.

Proof. The vectors $\mathbf{u}$ under consideration clearly form a module, in the sense of Appendix A. The result now follows from Lemma 1 and its corollary in Appendix A, since, as remarked above, the $\mathbf{u}$ are necessarily integral.

COROLLARY 1. *If $\rho_1, \ldots, \rho_s$ are real numbers and*

$$\rho_1\mathbf{u}^{(1)} + \ldots + \rho_s\mathbf{u}^{(s)} = 0,$$

then $$\rho_1 = \ldots = \rho_s = 0.$$

COROLLARY 2. *The $(n-1)$-dimensional vectors obtained from $\mathbf{u}^{(2)}, \ldots, \mathbf{u}^{(s)}$ by deleting the first element 0 have the same property with regard to the forms $L_2, \ldots, L_n$ as $\mathbf{u}^{(1)}, \ldots, \mathbf{u}^{(s)}$ have with regard to $L_1, \ldots, L_n$.*

Proofs. Clear.

LEMMA 4. *By appropriate choice† of $\mathbf{u}^{(1)}, \ldots, \mathbf{u}^{(s)}$ we may suppose that for any set of integers $w_1, \ldots, w_s$ there is a $\mathbf{z} \in \Lambda$ such that simultaneously* $$\mathbf{u}^{(t)}\mathbf{z} = w_t \quad (1 \leqslant t \leqslant s).$$

† This first phrase is unnecessary. It is not difficult to show that any $\mathbf{u}^{(1)}, \ldots, \mathbf{u}^{(s)}$ satisfying Lemma 3 also satisfy Lemma 4.

Proof. It is clearly enough to find vectors $\mathbf{z}^{(r)} \epsilon \Lambda$ $(1 \leqslant r \leqslant s)$ such that
$$\mathbf{u}^{(t)}\mathbf{z}^{(t)} = 1,$$
$$\mathbf{u}^{(t)}\mathbf{z}^{(r)} = 0 \quad \text{if} \quad r \neq t,$$

since then $\mathbf{z} = w_1 \mathbf{z}^{(1)} + \ldots + w_s \mathbf{z}^{(s)}$ does what is required.

Suppose first that $s = 1$. If d', d'' are two integers of the form $\mathbf{u}^{(1)}\mathbf{z}$, $\mathbf{z} \epsilon \Lambda$ then so is $a'd' + a''d''$ for any integers a', a''. The set of values taken by $\mathbf{u}^{(1)}\mathbf{z}$ is thus the set of all the multiples of some integer $d > 0$. But then $\mathbf{uz} = $ integer for all $\mathbf{z} \epsilon \Lambda$, where $\mathbf{u} = d^{-1}\mathbf{u}^{(1)}$. By Lemma 3, $d = 1$. Now $\mathbf{u}^{(1)}\mathbf{z}^{(1)} = d = 1$ for some $\mathbf{z}^{(1)} \epsilon \Lambda$ by the definition of d: which proves the lemma in this case.

Suppose that $s > 1$ and that Lemma 4 has been proved for smaller values of s. In particular, by considering only
$$L_2(\mathbf{x}), \ldots, L_n(\mathbf{x}),$$
we may find $\quad \mathbf{z}^{(r)} \quad (2 \leqslant r \leqslant s),$

such that $\quad \mathbf{u}^{(t)}\mathbf{z}^{(t)} = 1 \quad (2 \leqslant t \leqslant s),$
$$\mathbf{u}^{(t)}\mathbf{z}^{(r)} = 0 \quad (2 \leqslant r, t \leqslant s; r \neq t).$$

Denote the integers $\mathbf{u}^{(1)}\mathbf{z}^{(t)}$ by h_t. By considering
$$\mathbf{u}^{(1)} - h_2\mathbf{u}^{(2)} - \ldots - h_s\mathbf{u}^{(s)};$$
instead of $\mathbf{u}^{(1)}$ (which does not affect the truth of Lemma 3) we may suppose that
$$\mathbf{u}^{(1)}\mathbf{z}^{(t)} = 0 \quad (2 \leqslant t \leqslant s). \tag{4}$$

By the argument used when $s = 1$ we may certainly find a $\mathbf{z}^{(1)} \epsilon \Lambda$ such that
$$\mathbf{u}^{(1)}\mathbf{z}^{(1)} = 1. \tag{5}$$

Denote the integers $\mathbf{u}^{(t)}\mathbf{z}^{(1)}$ by g_t $(2 \leqslant t \leqslant s)$. By considering $\mathbf{z}^{(1)} - g_2\mathbf{z}^{(2)} - \ldots - g_s\mathbf{z}^{(s)}$ instead of $\mathbf{z}^{(1)}$ we may suppose, without upsetting (5), that
$$\mathbf{u}^{(t)}\mathbf{z}^{(1)} = 0 \quad (2 \leqslant t \leqslant s).$$

This proves Lemma 4.

COROLLARY. *If Theorem IV is true for all $\boldsymbol{\alpha}$ such that*
$$\mathbf{u}^{(t)}\boldsymbol{\alpha} = 0 \quad (1 \leqslant t \leqslant s), \tag{6}$$

then it is true universally.

Proof. If
$$\mathbf{u}^{(t)}\boldsymbol{\alpha} = w_t = \text{integer} \quad (1 \leqslant t \leqslant s),$$

and $\mathbf{z}' \in \Lambda$ is given by Lemma 4 then

$$\mathbf{u}^{(t)}\boldsymbol{\alpha}' = 0, \quad \boldsymbol{\alpha}' = \boldsymbol{\alpha} - \mathbf{z}' \quad (1 \leqslant t \leqslant s).$$

The truth of the theorem for $\boldsymbol{\alpha}'$ means that for any $\epsilon > 0$ there is a $\mathbf{z}'' \in \Lambda$ such that
$$|z_j'' - \alpha_j'| < \epsilon \quad (1 \leqslant j \leqslant n).$$

But then $\mathbf{z}^{(\epsilon)} = \mathbf{z}'' - \mathbf{z}' \in \Lambda$ and
$$|z_j^{(\epsilon)} - \alpha_j| < \epsilon \quad (1 \leqslant j \leqslant n),$$
as required.

LEMMA 5. *There is an $\epsilon_0 > 0$ such that all $\mathbf{z} \in \Lambda$ with*
$$|z_j| < \epsilon_0 \quad (1 \leqslant j \leqslant n),$$
satisfy $\mathbf{u}^{(t)}\mathbf{z} = 0 \,(1 \leqslant t \leqslant s)$.

Proof. We choose ϵ_0 so that
$$\epsilon_0(|u_{t1}| + \ldots + |u_{tn}|) < 1 \quad (1 \leqslant t \leqslant s).$$

If $\max |z_j| < \epsilon_0$ then $|\mathbf{u}^{(t)}\mathbf{z}| < 1$. But $\mathbf{u}^{(t)}\mathbf{z}$ is an integer if $\mathbf{z} \in \Lambda$.

LEMMA 6. *Suppose that there is an $\epsilon_1 > 0$ and a $\boldsymbol{\lambda} = (\lambda_1, \ldots, \lambda_n)$ such that all $\mathbf{z} \in \Lambda$ in*
$$|z_j| < \epsilon_1 \quad (1 \leqslant j \leqslant n) \tag{7}$$
also satisfy $\quad\quad\quad \boldsymbol{\lambda}\mathbf{z} = 0. \tag{8}$
Then $\quad\quad\quad \boldsymbol{\lambda} = \nu_1\mathbf{u}^{(1)} + \ldots + \nu_s\mathbf{u}^{(s)}, \tag{9}$
for some real $\nu_1, \ldots, \nu_s$.

Proof. Let ϵ be arbitrarily small, in particular
$$0 < \epsilon < \epsilon_1.$$

Suppose that $\mathbf{z} \neq 0$ belongs to Λ and satisfies (7). By Theorem VI of Chapter I there are integers $w \neq 0$, $\mathbf{t} = (t_1, \ldots, t_n)$ such that
$$\max_j |wz_j - t_j| < \epsilon < \epsilon_1. \tag{10}$$

But $w\mathbf{z} - \mathbf{t} \in \Lambda$ since $\mathbf{z} \in \Lambda$ and all points with integer co-ordinates are in Λ. Thus by the hypothesis of the lemma we have
$$\boldsymbol{\lambda}\mathbf{z} = \boldsymbol{\lambda}(w\mathbf{z} - \mathbf{t}) = 0,$$
and so $\quad\quad\quad\quad \boldsymbol{\lambda}\mathbf{t} = 0. \tag{11}$

By Lemma 2 there are numbers $\mu_1, \ldots, \mu_l$ ($l \leqslant n$) linearly independent over the rationals on which $\lambda_1, \ldots, \lambda_n$ are dependent, say

$$\boldsymbol{\lambda} = \mu_1 \mathbf{v}^{(1)} + \ldots + \mu_l \mathbf{v}^{(l)},$$

with rational vectors $\mathbf{v}^{(1)}, \ldots, \mathbf{v}^{(l)}$. Then (11) implies

$$\mu_1 \mathbf{v}^{(1)} \mathbf{t} + \ldots + \mu_l \mathbf{v}^{(l)} \mathbf{t} = 0,$$

and so

$$\mathbf{v}^{(i)} \mathbf{t} = 0 \quad (1 \leqslant i \leqslant l), \tag{12}$$

since $\mathbf{v}^{(i)}$ and $\mathbf{t}$ are rational. But now by (10) and (12)

$$\begin{aligned}
|\mathbf{v}^{(i)} \mathbf{z}| &\leqslant |w \mathbf{v}^{(i)} \mathbf{z}| \\
&= |\mathbf{v}^{(i)} (w\mathbf{z} - \mathbf{t})| \\
&< R^{(i)} \epsilon,
\end{aligned}$$

where $R^{(i)}$ is the sum of the absolute values of the co-ordinates of $\mathbf{v}^{(i)}$. Hence

$$\mathbf{v}^{(i)} \mathbf{z} = 0 \quad (1 \leqslant i \leqslant l), \tag{13}$$

since ϵ is arbitrarily small. Thus if (7) implies (8) it also implies the l equations (13) which are similar in form to the original one; and in which, moreover, all the elements of the $\mathbf{v}^{(i)}$ are rational. If the $\mathbf{v}^{(i)}$ are all of the form (9) then so is $\boldsymbol{\lambda}$. Hence it suffices to prove the lemma when all of $\lambda_1, \ldots, \lambda_n$ are rational.

Suppose, then, that $\lambda_1, \ldots, \lambda_n$ are all rational and let $\mathbf{z}$ be any vector of Λ. As before, there are integers $w \neq 0$, $\mathbf{t} = (t_1, \ldots, t_n)$ such that

$$|w z_j - t_j| < \epsilon_1 \quad (1 \leqslant j \leqslant n).$$

Since $w\mathbf{z} - \mathbf{t} \in \Lambda$ we have $\boldsymbol{\lambda}(w\mathbf{z} - \mathbf{t}) = 0$; so $\boldsymbol{\lambda}\mathbf{z}$ is rational since $\boldsymbol{\lambda}$, $\mathbf{t}$ are and $w \neq 0$. In particular all the coefficients of $\mathbf{x}$, $\mathbf{y}$ in

$$\sum_j \lambda_j (L_j(\mathbf{x}) - y_j)$$

are rational. There is thus an integer q such that all the coefficients in

$$\sum (q\lambda_j) (L_j(\mathbf{x}) - y_j)$$

are integers. By Lemma 3, $q\boldsymbol{\lambda} = p_1 \mathbf{u}^{(1)} + \ldots + p_s \mathbf{u}^{(s)}$ for some integers $p_1, \ldots, p_s$. But this is (9) with $\nu_i = p_i/q$.

LEMMA 7. *There are $n - s$ linearly independent† vectors*

$$\mathbf{z}^{(1)}, \ldots, \mathbf{z}^{(n-s)} \in \Lambda$$

in $\max |z_j| < \epsilon$, *for any given* $\epsilon > 0$.

† That is,

$\rho_1 \mathbf{z}^{(1)} + \ldots + \rho_{n-s} \mathbf{z}^{(n-s)} = 0$ for real $\rho_1, \ldots, \rho_{n-s}$ implies $\rho_1 = \ldots = \rho_{n-s} = 0$.

Proof. We construct $z^{(1)}, \ldots, z^{(n-s)}$ successively. Suppose we already have q vectors $z^{(1)}, \ldots, z^{(q)}$ where $0 \leqslant q < n-s$. There is clearly a λ such that

$$\lambda z^{(p)} = 0 \quad (1 \leqslant p \leqslant q), \tag{14}$$

which is not of the form $\lambda = \nu_1 u^{(1)} + \ldots + \nu_s u^{(s)}$. By Lemma 6 there is a $z^{(q+1)} \in \Lambda$ in $\max |z_j| < \epsilon$ such that $\lambda z^{(q+1)} \neq 0$. Then $z^{(1)}, \ldots, z^{(q+1)}$ are linearly independent vectors.

Proof of Theorem IV. By Lemma 4, Corollary, it is enough to consider α in the space $\mathscr{S}$ defined by

$$u^{(t)} \alpha = 0 \quad (1 \leqslant t \leqslant s).$$

Let $\epsilon > 0$ be arbitrarily small, and $z^{(1)}, \ldots, z^{(n-s)}$ the vectors of Lemma 7. By Lemma 5 they all lie in $\mathscr{S}$ if $\epsilon < \epsilon_0$ which we may suppose. Since $\mathscr{S}$ has dimension $n-s$ by Lemma 3, Corollary 1, we have

$$\alpha = \beta_1 z^{(1)} + \ldots + \beta_s z^{(s)},$$

for some real $\beta_1, \ldots, \beta_s$. There are integers $b_1, \ldots, b_s$ with

$$|\beta_r - b_r| \leqslant \tfrac{1}{2}.$$

Then
$$z^{(\epsilon)} = b_1 z^{(1)} + \ldots + b_s z^{(s)} \in \Lambda.$$

Further, the jth component of

$$\alpha - z^{(\epsilon)} = (\beta_1 - b_1) z^{(1)} + \ldots$$

has absolute value at most

$$|\beta_1 - b_1| \, |z_{1j}| + \ldots < \tfrac{1}{2} s \epsilon \leqslant \tfrac{1}{2} n \epsilon.$$

Since ϵ is arbitrarily small this proves the theorem.

We require later the

COROLLARY TO THEOREM IV.† *To each $\epsilon > 0$ there is an $X = X(\epsilon)$ such that for every real α satisfying B there is an integral a such that*

$$\| L_j(a) - \alpha_j \| < \epsilon \quad (1 \leqslant j \leqslant n)$$

and
$$\max(|a_1|, \ldots, |a_m|) \leqslant X(\epsilon).$$

Proof. We may suppose without loss of generality that $0 \leqslant \alpha_j < 1$. Then if α is in $\mathscr{S}$ the numbers $\beta_1, \ldots, \beta_{n-s}$ above are bounded and so there are only a finite number of possibilities for

---

† This also follows from the Heine-Borel theorem applied to the 'cube' $0 \leqslant \alpha_j \leqslant 1$.

$b_1, ..., b_{n-s}$, i.e. for **a**. The argument of Lemma 4, Corollary, shows that in general **a** may still be chosen out of a finite set since for $0 \leqslant \alpha_j < 1$ there are only a finite number of possibilities for $w_1, ..., w_s$. We may thus choose $X(\epsilon)$ to be the largest coordinate in the finite number of **x** required.

NOTES

§ 2. On letting $\epsilon \to 0$ in Theorem I B we see that there are infinitely many integer solutions x, y of (2·1) and (2·2) provided that there is no solution of $L_1 + \rho_1 = 0$. But if there is such a solution it may well be the only solution of (2·1) and (2·2), for example for $\rho_1 = \rho_2 = 0$, $L_1 L_2 = x^2 + xy - y^2$. Compare MORDELL (1951) and CASSELS (1954 b).

It was conjectured by Minkowski that if $L_j(\mathbf{x})$ are n linear forms in the n variables **x** with determinant $\Delta \neq 0$ and $\alpha_1, ..., \alpha_n$ are any numbers, then there are integers **x** with

$$\Pi \mid L_j(\mathbf{x}) - \alpha_j \mid \leqslant 2^{-n} \mid \Delta \mid.$$

Theorem I is a proof for $n = 2$. For the present state of knowledge, and a consideration of when infinitely many such integer **x** exist, see CASSELS (1952a) and ROGERS (1954).

For 'asymmetric' analogues of Theorem II see BLANEY (1950), SAWYER (1950) and BARNES & SWINNERTON-DYER (1955). For what happens when $x > 0$ see CASSELS (1954a).

Much work has been done on the best possible constant to replace $\frac{1}{4} \mid \Delta \mid$ in Theorem I if the product $L_1 L_2$ is a given indefinite quadratic form with integer coefficients, see BARNES & SWINNERTON-DYER (1952).

The use of continued fractions in the proof of Theorem II B exemplifies a common technique for both homogeneous and inhomogeneous problems. For a different proof of a rather weaker result see KANAGASABAPATHY (1952). For an even stronger result see BARNES (1956).

§ 3. KHINTCHINE (1926).

§ 5. There is a different (related) 'Kronecker's theorem' in PERRON (1913). For an interesting viewpoint see TURÁN (1953).

CHAPTER IV

UNIFORM DISTRIBUTION

1. Introduction. Suppose that θ is irrational. By the results of Chapter III there are integers q such that $\| q\theta - \alpha \|$ is arbitrarily small for any given α. In particular there are integers q such that $\{q\theta\}$ is arbitrarily close to any given number β in the unit interval $0 \leqslant \beta < 1$; or we may say that the numbers $\{q\theta\}$ are everywhere dense in the unit interval. Much more than this is true. Denote by $F_Q(\alpha, \beta)$ for $0 \leqslant \alpha < \beta \leqslant 1$ the number of integers q such that†

$$\alpha \leqslant \{q\theta\} < \beta, \quad 1 \leqslant q \leqslant Q.$$

Then the limit of $Q^{-1} F_Q(\alpha, \beta)$ is $\beta - \alpha$ as $Q \to \infty$; and indeed uniformly in α and β. This means that asymptotically each interval $\alpha \leqslant x < \beta$ contains 'the right number' of numbers $\{q\theta\}$.

Similar results hold with respect to simultaneous approximation. If $\theta_1, \ldots, \theta_n$ are such that no relation $u_1\theta_1 + \ldots + u_n\theta_n = v$ holds with integers $u_1, \ldots, u_n, v$ not all 0, then the sets of fractional parts

$$(\{q\theta_1\}, \ldots, \{q\theta_n\})$$

are uniformly distributed in the unit 'cube' $0 \leqslant x_j < 1$ in the obviously analogous way.

In § 2 we shall give a formal definition of uniform distribution, and in § 3 we shall give elementary deductions of the theorems referred to above (in a more general shape). Finally in §§ 4, 5 we shall discuss a very general criterion for uniform distribution using trigonometric sums, which gives very perspicuous proofs of the results of § 2 and of some general properties of uniform distribution.

It is convenient to introduce a new definition. We shall say that two numbers $z^{(1)}$, $z^{(2)}$ or, more generally, two vectors $\mathbf{z}^{(1)}$, $\mathbf{z}^{(2)}$ are CONGRUENT MODULO 1, or simply CONGRUENT, written

$$\mathbf{z}^{(1)} \equiv \mathbf{z}^{(2)},$$

---

† The reason for the appearance of one $<$ and one $\leqslant$ sign is purely technical.

if the difference $z^{(1)} - z^{(2)}$ has only integral co-ordinates. This definition is symmetrical in $z^{(1)}$ and $z^{(2)}$. Further, $z^{(1)} \equiv z^{(2)}$ and $z^{(2)} \equiv z^{(3)}$ imply $z^{(1)} \equiv z^{(3)}$. Vectors thus fall into classes of congruent ones. If we are interested merely in fractional parts, it is clearly irrelevant if we substitute congruent vectors for those originally considered. There is a representation of the real numbers in which congruent numbers are represented by the same point, got by, so to speak, rolling the real axis round a circle of unit perimeter. The number z is represented by a point on the circle at an angle $2\pi z$. In talking of the uniformity of distribution of the fractional parts of a set of numbers it is natural to have this representation in mind. Similarly the natural representation of the classes of congruent n-dimensional vectors is on an ' n-dimensional torus'; which, however, we shall not use.

2. Definition of discrepancy.

Suppose we are given a finite set of vectors
$$z^{(q)} = (z_{q1}, \ldots, z_{qn}) \quad (1 \leqslant q \leqslant Q)$$
in the unit 'cube'
$$0 \leqslant z_j < 1 \quad (1 \leqslant j \leqslant n). \tag{1}$$
Denote by $F(\boldsymbol{\alpha}, \boldsymbol{\beta})$, where $0 \leqslant \alpha_j < \beta_j \leqslant 1$ $(1 \leqslant j \leqslant n)$, the number of the $z^{(q)}$ lying in the parallelepiped
$$\alpha_j \leqslant z_j < \beta_j \tag{2}$$
of volume $\Pi(\beta_j - \alpha_j)$. Then
$$D = \sup_{\boldsymbol{\alpha}, \boldsymbol{\beta}} | Q^{-1} F(\boldsymbol{\alpha}, \boldsymbol{\beta}) - \Pi(\beta_j - \alpha_j) | \tag{3}$$
is the DISCREPANCY of the $z^{(q)}$. Clearly
$$0 < D \leqslant 1.$$
If we have an infinite sequence of vectors $z^{(q)}$ $(1 \leqslant q < \infty)$ in (1) let D_Q be the discrepancy of the first Q of them. If
$$D_Q \to 0 \tag{4}$$
as $Q \to \infty$ we say that the sequence is UNIFORMLY DISTRIBUTED in the unit cube. More generally, if we have a set of points $z^{(q)}$ in (1) labelled with sets of m positive integers $\mathbf{q} = (q_1, \ldots, q_m)$ let $D_{Q_1, \ldots, Q_m}$ be the discrepancy of the $Q_1 Q_2 \ldots Q_m$ vectors $z^{(q)}$ with

$1 \leqslant q_i \leqslant Q_i$ $(1 \leqslant i \leqslant m)$. We say that the $z^{(q)}$ are UNIFORMLY DISTRIBUTED if $D_{Q_1, \ldots, Q_m} \to 0$ as the $Q_i \to \infty$ independently.†

If z is any vector we shall denote by $\{z\}$ the vector of fractional parts $(\{z_1\}, \ldots, \{z_n\})$. We shall say that a set of vectors $z^{(q)}$ or $z^{(q)}$ is UNIFORMLY DISTRIBUTED MODULO 1 if the corresponding sets of fractional parts are uniformly distributed.

It would seem at first sight natural to measure the uniformity of distribution modulo 1 of a set of vectors $z^{(q)}$ $(1 \leqslant q \leqslant Q)$ by the discrepancy D of their fractional parts but it is technically convenient to do otherwise (cf. remarks at the end of § 1). Let $\Lambda^{(Q)}$ be the set of all $x = z^{(q)} + t$, t integral. For any α, β with $\beta_j \geqslant \alpha_j$ $(1 \leqslant j \leqslant n)$ let $F^*(\alpha, \beta)$ be the number of points $x \in \Lambda^{(Q)}$ in $\alpha_j \leqslant x_j < \beta_j$. Clearly

$$F^*(\alpha + t, \beta + t) = F^*(\alpha, \beta) \quad \text{(t integral)}, \tag{5}$$

and $$F^*(\alpha, \beta) = F(\alpha, \beta) \quad \text{for} \quad 0 \leqslant \alpha_j < \beta_j \leqslant 1, \tag{6}$$

where F is defined in terms of the fractional parts, as above. We define the DISCREPANCY MODULO 1 to be

$$D^* = \sup_{0 \leqslant \beta_j - \alpha_j \leqslant 1} | Q^{-1} F^*(\alpha, \beta) - \Pi(\beta_j - \alpha_j) |, \tag{7}$$

where α runs through all values, but may be restricted to the unit cube by (5). We shall show that

$$D \leqslant D^* \leqslant 2^n D, \tag{8}$$

the first half being trivial by (3), (6), (7). Any region $\alpha_j \leqslant x_j < \beta_j$ where $0 \leqslant \alpha_j < 1$, $\beta_j - \alpha_j \leqslant 1$ decomposes‡ into at most 2^n regions of the type $\alpha'_j \leqslant x_j < \beta'_j$, where for each j independently

$$\text{either} \quad 0 \leqslant \alpha'_j < \beta'_j \leqslant 1 \quad \text{or} \quad 1 \leqslant \alpha'_j < \beta'_j \leqslant 2. \tag{9}$$

Then $$F^*(\alpha, \beta) = \Sigma F^*(\alpha', \beta'),$$

by definition, and

$$\Pi(\beta_j - \alpha_j) = \Sigma \Pi(\beta'_j - \alpha'_j),$$

† That is, $D_{Q_1, \ldots, Q_m} < \epsilon$ whenever all Q_i are larger than a constant depending on ϵ.

‡ The reader is recommended to draw a diagram for, say, $n = 2$, $\alpha = (\frac{1}{4}, \frac{3}{4})$, $\beta = (\frac{3}{4}, \frac{3}{2})$.

because $\Pi(\beta_j - \alpha_j)$ is the volume of the whole region and $\Pi(\beta'_j - \alpha'_j)$ that of one of the parts. Hence

$$|Q^{-1}F^*(\alpha, \beta) - \Pi(\beta_j - \alpha_j)|$$
$$\leqslant \Sigma |Q^{-1}F^*(\alpha', \beta') - \Pi(\beta'_j - \alpha'_j)|. \qquad (10)$$

But each of the summands in (10) is at most D by (2), (3), (5), (6), (9). Since there are at most 2^n summands this completes the proof of (8), by (7).

By (8) uniform distribution modulo 1 may equally well be defined by $D_Q^* \to 0$ (in an obvious notation) as by $D_Q \to 0$.

For any fixed γ ($\gamma_j > 0$) the average of $F^*(\alpha, \alpha + \gamma)$ over α is what one would expect:

LEMMA 1. *For any γ with $\gamma_j > 0$ ($1 \leqslant j \leqslant n$) not necessarily $\gamma_j \leqslant 1$,*

$$\int \ldots \int_{0 \leqslant \alpha_j < 1} F^*(\alpha, \alpha + \gamma)\, d\alpha = Q\Pi\gamma_j; \quad d\alpha = d\alpha_1 \ldots d\alpha_n.$$

Proof. $$F^*(\alpha, \alpha + \gamma) = \sum_{1 \leqslant q \leqslant Q} f_q(\alpha, \gamma), \qquad (11)$$

where $f_q(\alpha, \gamma)$ is the number of vectors $\mathbf{x} = \mathbf{z}^{(q)} + \mathbf{t}$ ($\mathbf{t}$ integral) in $\alpha_j \leqslant x_j < \alpha_j + \gamma_j$ ($1 \leqslant j \leqslant n$). But

$$f_q(\alpha, \gamma) = \sum_{\mathbf{t}} \phi_q(\alpha - \mathbf{t}, \gamma),$$

where the sum is over all integer vectors $\mathbf{t}$ and $\phi_q(\beta, \gamma) = 1$ if β is in the region $$z_{qj} \geqslant \beta_j > z_{qj} - \gamma_j \quad (1 \leqslant j \leqslant n),$$

of volume $\Pi\gamma_j$; but $\phi_q(\beta, \gamma) = 0$ otherwise. Hence

$$\int \ldots \int_{0 \leqslant \alpha_j < 1} f_q(\alpha, \gamma)\, d\alpha$$
$$= \sum_{\mathbf{t}} \int \ldots \int_{0 \leqslant \alpha_j < 1} \phi_q(\alpha - \mathbf{t}, \gamma)\, d\alpha$$
$$= \int \ldots \int_{-\infty < \alpha_j < \infty} \phi_q(\alpha, \gamma)\, d\alpha$$
$$= \Pi\gamma_j.$$

This proves the lemma by (11).

3. Uniform distribution for linear forms.

THEOREM I. *Let* $L_j(\mathbf{x})$ *for* $1 \leqslant j \leqslant n$ *be homogeneous forms in the* m *variables* $\mathbf{x} = (x_1, \ldots, x_m)$. *Suppose that the only set of integers* $u_1, \ldots, u_n$ *such that*
$$u_1 L_1(\mathbf{x}) + \ldots + u_n L_n(\mathbf{x})$$
has integer coefficients in $x_1, \ldots, x_m$ *is* $u_1 = \ldots = u_n = 0$. *Then the set of vectors* $\mathbf{z}^{(\mathbf{x})} = (L_1(\mathbf{x}), \ldots, L_n(\mathbf{x}))$ *for integral* $\mathbf{x}$ *is uniformly distributed modulo* 1.

Note. The reader will have no difficulty in enunciating the corresponding result when $\Sigma u_i L_i(\mathbf{x})$ has integer coefficients for some $\mathbf{u} \neq \mathbf{0}$ and in modifying the proof to fit.

Proof. The principle is quite clear when $m = n = 1$. To avoid an unnecessary proliferation of indexes we confine ourselves to this. Then $L_1(\mathbf{x}) = \theta x_1$ for some irrational number θ. On writing x for x_1, we have to show that θx ($x = 1, 2, \ldots$) is uniformly distributed modulo 1. By Theorem IV, Corollary, of Chapter III on p. 58, for any $\epsilon > 0$ there is an $X(\epsilon)$ such that for any real α there are integers x, y with

$$|\theta x - y - \alpha| < \epsilon, \quad |x| \leqslant X(\epsilon). \tag{1}$$

Now let $Q > X(\epsilon)$ and consider $\Lambda^{(Q)}$ of the last section, i.e. the set of numbers $q\theta - p$ where q, p are integers and $1 \leqslant q \leqslant Q$. Let $F^*(\alpha, \beta)$ be defined with respect to $\Lambda^{(Q)}$ as in § 2. We shall show that

$$F^*(\alpha, \beta) \leqslant F^*(\gamma, \delta) + X, \quad X = X(\epsilon); \tag{2}$$

where $\alpha, \beta, \gamma, \delta$ are any four numbers satisfying

$$0 < \beta - \alpha = \delta - \gamma - 2\epsilon \leqslant 1. \tag{3}$$

Here $F^*(\alpha, \beta)$ is the number of pairs of integers p, q with

$$1 \leqslant q \leqslant Q \tag{4}$$

and
$$\alpha \leqslant q\theta - p < \beta. \tag{5}$$

Let x_0, y_0 be a solution of

$$|x_0 \theta - y_0 - (\gamma + \epsilon - \alpha)| < \epsilon, \quad |x_0| \leqslant X(\epsilon). \tag{6}$$

Then $q' = q + x_0$, $p' = p + y_0$ gives an integer solution of

$$\gamma \leqslant q'\theta - p' < \delta \tag{7}$$

by (3), (5) and (6). But $F^*(\gamma, \delta)$ is the number of solutions of (7) with

$$1 \leqslant q' \leqslant Q. \tag{8}$$

Since $1 \leqslant q \leqslant Q$, the inequality (8) holds except when $1 \leqslant q \leqslant |x_0|$ if $x_0 < 0$ and except when $Q - x_0 + 1 \leqslant q \leqslant Q$ if $x_0 > 0$. In any case there are at most $|x_0| \leqslant X$ values of q such that (4) but not (8) holds; and for each q there is at most one p in (5) since $\beta \leqslant \alpha + 1$. Hence the number of solutions of (4), (5) is at most X more than the number of solutions of (7), (8). This is just (2).

With fixed α, β we have thus for $0 < \beta - \alpha \leqslant 1$

$$F^*(\alpha, \beta) \leqslant X + \int_0^1 F^*(\gamma, \gamma + \beta - \alpha + 2\epsilon)\, d\gamma$$
$$= X + (\beta - \alpha + 2\epsilon)\, Q, \tag{9}$$

by (2), (3) and Lemma 1. Similarly, on interchanging the pairs α, β and γ, δ we have for $2\epsilon < \beta - \alpha \leqslant 1$

$$F^*(\alpha, \beta) \geqslant -X + \int_0^1 F^*(\gamma, \gamma + \beta - \alpha - 2\epsilon)\, d\gamma$$
$$= -X + (\beta - \alpha - 2\epsilon)\, Q. \tag{10}$$

Finally, trivially,

$$F^*(\alpha, \beta) \geqslant 0 \text{ always.} \tag{11}$$

Hence, by (9), (10), (11) we have

$$|Q^{-1} F^*(\alpha, \beta) - (\beta - \alpha)| \leqslant 2\epsilon + Q^{-1} X < 3\epsilon,$$

if $Q > \epsilon^{-1} X(\epsilon) = Q_0(\epsilon)$ (say). Hence, by definition, $D_Q^* < 3\epsilon$ if $Q > Q_0(\epsilon)$. Since ϵ is arbitrary, this completes the proof.

4. Weyl's criteria. For simplicity of notation we shall speak only of a sequence of vectors $\mathbf{z}^{(q)}$ $(q = 1, 2, \ldots)$. The generalization to $\mathbf{z}^{(\mathbf{q})}$, $\mathbf{q} = (q_1, \ldots, q_m)$ is immediate. We shall prove two theorems of Weyl.

THEOREM II. *Let $\mathbf{z}^{(q)}$ $(q = 1, 2, \ldots)$ be a sequence of vectors in the n-dimensional unit cube $0 \leqslant z_j < 1$. The necessary and sufficient condition that it be uniformly distributed in the unit cube is that*

$$\lim_{Q \to \infty} Q^{-1} \sum_{q \leqslant Q} f(\mathbf{z}^{(q)}) = \int \cdots \int_{0 \leqslant z_j < 1} f(\mathbf{z})\, d\mathbf{z}, \tag{1}$$

for all real or complex Riemann-integrable functions $f(\mathbf{z})$ defined in the unit cube.

5

THEOREM III. *Let* $z^{(q)}$ ($q = 1, 2, \ldots$) *be any sequence of n-dimensional vectors, not necessarily restricted to lie in the unit cube. The necessary and sufficient condition that it be uniformly distributed modulo* 1 *is that*

$$\lim_{Q \to \infty} Q^{-1} \sum_{q \leqslant Q} e(\mathbf{t} z^{(q)}) = 0 \qquad (2)$$

for all integral vectors $\mathbf{t} \neq \mathbf{0}$, *where*

$$e(x) = e^{2\pi i x}, \quad i^2 = -1.$$

Note 1. There is no talk of uniformity in (2). It is asserted only that (2) holds for each $\mathbf{t} \neq \mathbf{0}$.

Note 2. Suppose in particular that $z^{(q)} = q\boldsymbol{\theta}$, where

$$\boldsymbol{\theta} = (\theta_1, \ldots, \theta_n),$$

and there is no relation $\mathbf{t}\boldsymbol{\theta} = $ integer with integral $\mathbf{t} \neq \mathbf{0}$. Then $e(\mathbf{t}\boldsymbol{\theta}) \neq 1$ for each $\mathbf{t}$ and

$$\left| Q^{-1} \sum_{q \leqslant Q} e(\mathbf{t} z^{(q)}) \right| = \left| Q^{-1} \sum_{q \leqslant Q} e(q\,\mathbf{t}\boldsymbol{\theta}) \right|$$

$$= \left| \frac{e(\mathbf{t}\boldsymbol{\theta})\,(1 - e(Q\mathbf{t}\boldsymbol{\theta}))}{Q(1 - e(\mathbf{t}\boldsymbol{\theta}))} \right|$$

$$\leqslant \frac{2}{Q\,|\,1 - e(\mathbf{t}\boldsymbol{\theta})\,|} \to 0.$$

Hence Theorem III applies and the $z^{(q)}$ are uniformly distributed modulo 1. This is the special case $m = 1$ of Theorem I. The general Theorem I follows similarly from the appropriate generalization of Theorem III.

Note 3. Since neither of the statements in Theorem III is affected by replacing the $z^{(q)}$ by congruent vectors modulo 1, we may suppose, if we like, that they all lie in the unit cube $0 \leqslant z_j < 1$, in which case uniform distribution modulo 1 becomes uniform distribution.

Proof of Theorems II, III. To simplify the notation we assume that $n = 1$ since there are no great additional complications when $n > 1$. Our vectors $z^{(q)}$ are thus substantially numbers, which we shall denote by $z^{(q)}$. To prove Theorems II, III (and rather more) simultaneously it is enough to prove the cycle of implications

$$A \to B \to C \to D \to A$$

about $z^{(q)}$, where

$$0 \leqslant z^{(q)} < 1 \quad (q = 1, 2, \ldots), \tag{3}$$

and A, B, C, D are as follows:

Statement A. $z^{(q)}$ is uniformly distributed in $0 \leqslant z < 1$.

Statement B.

$$Q^{-1} F_Q(\alpha, \beta) \to \beta - \alpha \quad (Q \to \infty), \tag{4}$$

for each pair of numbers α, β, $0 \leqslant \alpha < \beta \leqslant 1$, where (as before) $F_Q(\alpha, \beta)$ is the number of solutions of

$$\alpha \leqslant z^{(q)} < \beta, \quad 1 \leqslant q \leqslant Q. \tag{5}$$

Uniformity with respect to α and β is not assumed.

Statement C.

$$Q^{-1} \sum_{q \leqslant Q} f(z^{(q)}) \to \int_0^1 f(z) \, dz, \tag{6}$$

for all functions $f(z)$ Riemann-integrable in $0 \leqslant z \leqslant 1$.

Statement D.
$$Q^{-1} \sum_{q \leqslant Q} e(t z^{(q)}) \to 0 \tag{7}$$

for all integers $t \neq 0$. Again, no uniformity with respect to t is assumed.

Proof that A *implies* B. Trivial since B is an ostensibly† weaker form of A.

Proof that C *implies* D. Again trivial since $e(tz)$ is Riemann-integrable (indeed continuous) and

$$\int_0^1 e(tz) \, dz = 0 \quad (t \neq 0, \text{ integer}).$$

Proof that B *implies‡* C. By considering the real and imaginary parts of $f(z)$ separately we may suppose without loss of generality that $f(z)$ is real and, by adding an appropriate constant to $f(z)$, that
$$f(z) \geqslant 0. \tag{8}$$

Let $\epsilon > 0$ be given and let V be a large positive integer. For all

---

† The reader might find it interesting to construct a simple direct proof of the converse implication.

‡ The converse implication C → B is trivial on considering the function $f(z)$ defined by $f(z) = 1$ if $\alpha \leqslant z < \beta$ but vanishing elsewhere.

integers v, $1 \leqslant v \leqslant V$, let m_v, M_v be respectively the minimum and the maximum of $f(z)$ in

$$v - 1 \leqslant Vz < v, \qquad (9)$$

so that

$$0 \leqslant m_v \leqslant M_v. \qquad (10)$$

Then, since $f(z)$ is Riemann-integrable, we have

$$\int_0^1 f(z)\, dz - \epsilon \leqslant V^{-1} \Sigma m_v \leqslant V^{-1} \Sigma M_v \leqslant \int_0^1 f(z)\, dz + \epsilon, \qquad (11)$$

if V is large enough. Let now V be some fixed integer such that (11) holds. The number $F_Q\left(\dfrac{v-1}{V}, \dfrac{v}{V}\right) = \phi_v$ (say) of $z^{(q)}$ in (9) with $1 \leqslant q \leqslant Q$ satisfies

$$(1 - \epsilon)\, V^{-1} \leqslant Q^{-1} \phi_v \leqslant (1 + \epsilon)\, V^{-1}$$

for all large enough Q, by the hypothesis that B is true. For such Q we have (using (10))

$$Q^{-1} \sum_{q \leqslant Q} f(z^{(q)}) = \sum_v Q^{-1} \sum_{\substack{q \leqslant Q \\ v-1 \leqslant Vz^{(q)} < v}} f(z^{(q)})$$

$$\leqslant \sum_v Q^{-1} \phi_v M_v$$

$$\leqslant (1 + \epsilon)\, V^{-1} \Sigma M_v$$

$$\leqslant (1 + \epsilon) \left(\int_0^1 f(z)\, dz + \epsilon \right)$$

by (11). Similarly

$$Q^{-1} \sum_{q \leqslant Q} f(z^{(q)}) \geqslant (1 - \epsilon) \left(\int_0^1 f(z)\, dz - \epsilon \right).$$

Since ϵ is arbitrarily small, this proves (6).

Proof that D *implies* A.

LEMMA 2. *To every $\epsilon > 0$ there is an $E = E(\epsilon)$ with the following property:*

For each α, β, $0 \leqslant \alpha < \beta \leqslant 1$ there are functions $f_-(z), f_+(z)$ periodic modulo 1 with continuous second derivatives such that

(i) $0 \leqslant f_-(z) \leqslant 1$, $\quad 0 \leqslant f_+(z) \leqslant 1$.

(ii) $f_+(z) = 1$ *if* $\alpha \leqslant z < \beta$,
$\quad f_-(z) = 0$ *if* $0 \leqslant z < \alpha$ *or* $\beta \leqslant z < 1$.

(iii) $\int_0^1 f_+(z)\,dz \leqslant \beta - \alpha + \epsilon,$

$\int_0^1 f_-(z)\,dz \geqslant \beta - \alpha - \epsilon.$

(iv) $|f_+''(z)| \leqslant E,\ \ |f_-''(z)| \leqslant E$ *for all* z.

Note. It is clear from (i) and (ii) that

$$\sum_{q \leqslant Q} f_-(z^{(q)}) \leqslant F_Q(\alpha, \beta) \leqslant \sum_{q \leqslant Q} f_+(z^{(q)}). \qquad (12)$$

As we shall see, (iii), (iv) ensure that $f_\pm(z)$ have convenient Fourier expansions.

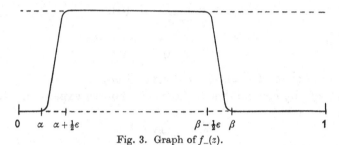

Fig. 3. Graph of $f_-(z)$.

Proof. We first construct $f_-(z)$. If $\beta - \alpha \leqslant \epsilon$ the function $f_-(z) = 0$ (all z) does what is required, so we may suppose that $\beta > \alpha + \epsilon$. There is certainly a twice differentiable function $\phi(x)$ defined in $0 \leqslant x \leqslant 1$ with the following properties

$$\phi(0) = \phi'(0) = \phi''(0) = 0;$$
$$\phi(1) = 1, \quad \phi'(1) = \phi''(1) = 0;$$
$$0 \leqslant \phi(x) \leqslant 1 \quad \text{for} \quad 0 \leqslant x \leqslant 1:$$

for example

$$\phi(x) = \begin{cases} 8x^3 - 8x^4 & \text{if } 0 \leqslant x \leqslant \tfrac{1}{2} \\ 1 - 8(1-x)^3 + 8(1-x)^4 & \text{if } \tfrac{1}{2} \leqslant x \leqslant 1 \end{cases}$$

will do, since $\phi'(\tfrac{1}{2})$, $\phi''(\tfrac{1}{2})$ exist and clearly $\phi(x)$ has the other properties required. We define as in Figure 3

$$f_-(z) = \begin{cases} 0 & \text{if } 0 \leqslant z < \alpha, \\ \phi(2\epsilon^{-1}(z-\alpha)) & \text{if } \alpha \leqslant z < \alpha + \tfrac{1}{2}\epsilon, \\ 1 & \text{if } \alpha + \tfrac{1}{2}\epsilon \leqslant z < \beta - \tfrac{1}{2}\epsilon, \\ \phi(2\epsilon^{-1}(\beta - z)) & \text{if } \beta - \tfrac{1}{2}\epsilon \leqslant z < \beta, \\ 0 & \text{if } \beta \leqslant z < 1. \end{cases}$$

Clearly $f_-(z)$ is twice differentiable and

$$|f''_-(z)| \leqslant 4\epsilon^{-2} \max |\phi''(x)| = E(\epsilon),$$

say, where E is independent of α and β. Thus $f_-(z)$ satisfies (i), (ii) and (iv). It also satisfies (iii) since

$$\int_0^1 f_-(z)\, dz \geqslant \int_{\alpha+\frac{1}{2}\epsilon}^{\beta-\frac{1}{2}\epsilon} dz = \beta - \alpha - \epsilon.$$

The construction of $f_+(z)$ is similar.

COROLLARY. $\qquad f_+(z) = \sum_{-\infty < t < \infty} c_t^+ e(tz),$ $\qquad\qquad$ (13)

$$f_-(z) = \sum_{-\infty < t < \infty} c_t^- e(tz), \qquad\qquad (14)$$

where $\qquad\qquad c_0^- \geqslant \beta - \alpha - \epsilon, \quad c_0^+ \leqslant \beta - \alpha + \epsilon,$ $\qquad\qquad$ (15)

$$|c_t^\pm| \leqslant t^{-2} M \quad (t \neq 0), \qquad\qquad (16)$$

for an M which depends on ϵ but not on α or β.

Proof. By general theory there exist Fourier expansions (13), (14), where

$$c_0^\pm = \int_0^1 f_\pm(z)\, dz$$

and, for $t \neq 0$,

$$c_t^\pm = \int_0^1 f_\pm(z)\, e(-tz)\, dz$$

$$= \frac{-1}{4\pi^2 t^2} \int_0^1 f''_\pm(z)\, e(-tz)\, dz,$$

on integrating by parts twice. The corollary now follows from (iii), (iv).

The proof that D implies A is now immediate. Let $\epsilon > 0$ be arbitrarily small and construct $f_\pm(z)$ for any $\alpha, \beta, 0 \leqslant \alpha < \beta \leqslant 1$ by Lemma 2. Then by (12) and Lemma 2, Corollary, we have

$$F_Q(\alpha, \beta) \leqslant \sum_{q \leqslant Q} f_+(z^{(q)})$$

$$= \sum_{-\infty < t < \infty} c_t^+ \sum_{q \leqslant Q} e(tz^{(q)})$$

$$\leqslant Q(\beta - \alpha + \epsilon) + \sum_{t \neq 0} t^{-2} M \Big| \sum_{q \leqslant Q} e(tz^{(q)}) \Big|.$$

There is a similar estimate from below, using $f_-(z)$, and so

$$D_Q \leqslant \epsilon + M Q^{-1} \sum_{t \neq 0} t^{-2} \Big| \sum_{q \leqslant Q} e(tz^{(q)}) \Big|. \qquad\qquad (17)$$

Now, since Σt^{-2} is convergent, we can choose a T such that $M \sum_{t \geqslant T} t^{-2} < \epsilon$. Hence, since trivially the inner sum in (17) is $\leqslant Q$, the contribution of the terms in (17) with $|t| \geqslant T$ is at most 2ϵ. We keep now ϵ, T fixed. If D is true, we have

$$\left| Q^{-1} \sum_{q \leqslant Q} e(tz^{(q)}) \right| < (TM)^{-1} \epsilon \tag{18}$$

for all t in $0 < |t| \leqslant T$ and all $Q \geqslant$ some $Q_0(\epsilon)$, since there are only a finite number of t in consideration. Hence $D_Q < 5\epsilon$ for all $Q \geqslant Q_0(\epsilon)$ by (17), (18). Since ϵ is arbitrarily small, this proves A.

5. Consequences of Weyl's criteria.

THEOREM IV. *The necessary and sufficient condition that the sequence of 2-dimensional vectors $\mathbf{z}^{(q)} = (x^{(q)}, y^{(q)})$ be uniformly distributed modulo 1 is that the 1-dimensional sequences $ux^{(q)} + vy^{(q)}$ shall be uniformly distributed modulo 1 for all pairs of integers other than $u = v = 0$.*

THEOREM V. *A sufficient condition† that the 1-dimensional sequence $z^{(q)}$ be uniformly distributed modulo 1 is that the sequence $z^{(q+h)} - z^{(q)}$ be uniformly distributed for each integer $h > 0$.*

THEOREM VI. *Let the polynomial*

$$f(x) = \alpha_r x^r + \ldots + \alpha_0 \tag{1}$$

have at least one irrational coefficient α_j with $j > 0$. Then the sequence

$$z^{(q)} = f(q) \tag{2}$$

is uniformly distributed modulo 1.

Theorem IV is an immediate consequence of Theorem III so we omit a formal proof. We shall deduce Theorem VI from Theorem V, which is an almost immediate consequence of Theorem III and

LEMMA 3. *Let $u_1, \ldots, u_Q$ be any (real or) complex numbers with conjugates $\bar{u}_q$ and let $1 \leqslant H \leqslant Q$. Then*

$$H^2 \left| \sum_{1 \leqslant q \leqslant Q} u_q \right|^2 \leqslant H(H + Q - 1) \sum_{1 \leqslant q \leqslant Q} |u_q|^2$$
$$+ 2(H + Q - 1) \sum_{0 < h < H} (H - h) \left| \sum_{1 \leqslant q \leqslant Q - h} \bar{u}_q u_{q+h} \right|. \tag{3}$$

† Not necessary. Gegenbeispiel: $z^{(q)} = q\theta$ with irrational θ.

Proof. It is convenient temporarily to use the convention

$$u_q = 0 \quad \text{if} \quad q \leqslant 0 \quad \text{or} \quad q > Q.$$

Then
$$H \sum_{1 \leqslant q \leqslant Q} u_q = \sum_{0 < p < H+Q} \left(\sum_{0 \leqslant r < H} u_{p-r} \right).$$

Hence, by Schwarz's inequality† the left-hand side of (3) is at most $H + Q - 1$ times

$$\sum_{0 < p < H+Q} \left| \sum_{0 \leqslant r < H} u_{p-r} \right|^2$$
$$= \sum_{\substack{0 < p < H+Q \\ 0 \leqslant r, s < H}} u_{p-r} \bar{u}_{p-s}. \tag{4}$$

But any term $|u_q|^2$ occurs in (4) precisely H times, namely for $q = p - r = p - s$ and $0 \leqslant r < H$. Any term $u_q \bar{u}_{q+h}$ or $\bar{u}_q u_{q+h}$ $(h > 0)$ can occur only if $h < H$ and then it occurs precisely $H - h$ times. Hence, (4) is

$$H \sum_{1 \leqslant q \leqslant Q} |u_q|^2 + \sum_{0 < h < H} (H-h) \sum_{1 \leqslant q \leqslant Q-h} (u_q \bar{u}_{q+h} + \bar{u}_q u_{q+h}),$$

and the lemma follows.

COROLLARY. *Suppose that*

$$Q^{-1} \sum_{1 \leqslant q \leqslant Q} e(z^{(q+h)} - z^{(q)}) \to 0 \quad (Q \to \infty)$$

for each $h > 0$ (not necessarily uniformly in h). Then

$$Q^{-1} \sum_{1 \leqslant q \leqslant Q} e(z^{(q)}) \to 0 \quad (Q \to \infty).$$

Proof. Put $u_q = e(z^{(q)})$. For all $Q > H > 0$ we have

$$Q^{-2} \left| \sum_{1 \leqslant q \leqslant Q} e(z^{(q)}) \right|^2 \leqslant \frac{H+Q-1}{HQ}$$
$$+ 2 \sum_{0 < h < H} \frac{(H+Q-1)(H-h)}{H^2 Q^2} \left| \sum_{1 \leqslant q \leqslant Q-h} e(z^{(q+h)} - z^{(q)}) \right|. \tag{5}$$

If, now, H is fixed and $Q \to \infty$ the right-hand side of (5) tends to H^{-1}; which is arbitrarily small by appropriate initial choice of H. Hence the left-hand side of (5) must tend to 0 as $Q \to \infty$.

† Namely that $|\Sigma \eta_l \zeta_l|^2 \leqslant \Sigma |\zeta_l|^2 \Sigma |\eta_l|^2$ for all sets of complex numbers $\eta_l, \zeta_l \, (1 \leqslant l \leqslant L)$. For a proof, cf. footnote p. 122.

Proof of Theorem V. By the hypothesis that $z^{(q+h)} - z^{(q)}$ is uniformly distributed and Theorem III,

$$Q^{-1} \sum_{1 \leqslant q \leqslant Q} e(t(z^{(q+h)} - z^{(q)})) \to 0$$

for all integers $h > 0$, $t \neq 0$. By Lemma 3, Corollary, applied to $tz^{(q)}$ we deduce

$$Q^{-1} \sum_{1 \leqslant q \leqslant Q} e(tz^{(q)}) \to 0 \quad (t \neq 0).$$

The $z^{(q)}$ are thus uniformly distributed modulo 1, by Theorem III again.

Proof of Theorem VI. Suppose, first, that the leading coefficient α_r is irrational. When $r = 1$ Theorem VI is a special case of Theorem I, so we may assume that $r > 1$ and that the result has been proved for $r - 1$. For any fixed integer $h > 0$ we have

$$z^{(q+h)} - z^{(q)} = f(q+h) - f(q),$$

a polynomial in q of degree $r - 1$ with irrational leading coefficient $h\alpha_r$. Hence the result for r follows from that for $r - 1$ and Theorem V.

If, however, α_r is rational there is some s, $0 < s < r$, such that α_s is irrational but $\alpha_{s+1}, ..., \alpha_r$ are rational. Let $M > 0$ be an integer such that $M\alpha_{s+1}, ..., M\alpha_r$ are integers. It is clearly enough to show that

$$z^{(Mq+m)} = \zeta^{(q)} \quad (q = 1, 2, ...),$$

is uniformly distributed modulo 1 for each of $m = 0, 1, ..., M-1$. But

$$\begin{aligned}
\zeta^{(q)} &= \alpha_0 + \alpha_1(Mq+m) + ... + \alpha_r(Mq+m)^r \\
&\equiv \alpha_0 + \alpha_1(Mq+m) + ... + \alpha_s(Mq+m)^s \\
&\quad + \alpha_{s+1} m^{s+1} + ... + \alpha_r m^r \quad \text{(modulo 1)}, \\
&= \beta_0 + \beta_1 q + ... + \beta_s q^s,
\end{aligned}$$

where $\beta_1, ..., \beta_s$ are independent of q. In particular $\beta_s = M^s \alpha_s$ is irrational. This is the first case, so the theorem is proved generally.

NOTES

§ 2. A striking theorem of Aardenne-Ehrenfest states that $QD_Q \to \infty$ for any sequence of real numbers in $0 \leqslant z < 1$ whatsoever, see ROTH (1954).

§ 3. As in § 3 of Chapter III it can be shown that Theorem I cannot be replaced by any quantitative formulation however weak. For work on special θ when $m = n = 1$ see KOKSMA (1936), Kap. IX, § 2. For example $QD_Q = O(\log Q)$ if θ is a quadratic irrational.

§ 4. For a simple quantitative form of Theorem III which is also the strongest to date, see ERDÖS & TURÁN (1948).

The $e(\mathbf{tz})$ for integral $\mathbf{t}$ are, of course, the characters of the additive group of vectors modulo 1. For a proof of a rather weak result using the theory of topological groups see PONTRJAGIN (1938), § 33.

§ 5. For an interesting discussion see VAN DER CORPUT (1931). For a quantitative form of Theorem V see CASSELS (1953).

Much quantitative work has been done on the distribution of fractional parts of polynomials and the associated problem of estimating $\Sigma e(f(q))$, where f is a polynomial: see VINOGRADOFF (1947).

CHAPTER V

TRANSFERENCE THEOREMS

1. Introduction. In this chapter we show how information about some problem for a given set of linear forms sometimes gives information about another problem for a related set of linear forms. Let

$$L_j(\mathbf{x}) = \sum_i \theta_{ji} x_i \quad (1 \leqslant i \leqslant m,\ 1 \leqslant j \leqslant n)$$

be n linear forms in m variables and let

$$M_i(\mathbf{u}) = \sum_j \theta_{ji} u_j$$

be the TRANSPOSED set of m linear forms in n variables. By Theorem VI of Chapter I there is always an integral $\mathbf{x} \neq \mathbf{0}$ such that

$$\| L_j(\mathbf{x}) \| \leqslant C \ (1 \leqslant j \leqslant n), \quad | x_i | \leqslant X \ (1 \leqslant i \leqslant m) \tag{1}$$

for any $X > 1$ and $C = X^{-m/n}$. In § 2 we shall show that if (1) is soluble with $\mathbf{x} \neq \mathbf{0}$ for some X and some C much smaller than $X^{-m/n}$ then the transposed set of inequalities

$$\| M_i(\mathbf{u}) \| \leqslant D, \quad | u_j | \leqslant U, \tag{2}$$

is soluble with $\mathbf{u} \neq \mathbf{0}$ for some U depending on X and C and some D much smaller than the expected $U^{-n/m}$. The particular case $m = 1$ of this result sets up a relation between the problem of simultaneous approximation to n irrationals

$$\| \theta_j x \| \leqslant C, \quad | x | \leqslant X, \tag{3}$$

where $\theta_j = \theta_{j1}$ and $x = x_1$, and the approximation of a single linear form,

$$\| \theta_1 u_1 + \ldots + \theta_n u_n \| \leqslant D, \quad | u_j | \leqslant U. \tag{4}$$

This case will be discussed more minutely in § 3.

In §§ 4, 5 we shall show that there is a relation between the 'homogeneous' problem (1) and the corresponding 'inhomogeneous' problem of solving

$$\| L_j(\mathbf{x}) - \alpha_j \| \leqslant C_1, \quad | x_i | \leqslant X_1, \tag{5}$$

with integral $\mathbf{x}$ for given $\boldsymbol{\alpha}$. Roughly speaking, we shall show that if $L_1, \ldots, L_n$ approximate well simultaneously to 0 (i.e. if (1) is soluble for some X with an unusually small C) then there is an $\boldsymbol{\alpha}$ which is badly approximated to by $L_1, \ldots, L_n$ (i.e. there is an X_1, depending on X and C, such that (5) is soluble only with an abnormally large value of C_1); and conversely. In §§ 6, 7 we use the theory to clear up some points left in Chapter III.

On putting the results of §§ 2, 4 together, we see that the homogeneous problems for $L_j(\mathbf{x})$, $M_i(\mathbf{u})$ and the corresponding inhomogeneous problems each give information about the other. In particular, both a necessary and a sufficient condition that (5) should be soluble for all $\boldsymbol{\alpha}$ for some given C_1, X_1 is that (2) should be insoluble for some D and U depending only on C_1, X_1: of course D and U are not the same for the necessary and for the sufficient condition. The special case of Kronecker's theorem (Theorem IV of Chapter III) in which $u_1 L_1(\mathbf{x}) + \ldots + u_n L_n(\mathbf{x})$ is not a form with integer coefficients in $\mathbf{x}$ for any integers $\mathbf{u} \neq \mathbf{0}$, may be regarded as a 'limiting case' $C_1 = \epsilon > 0$, $X_1 = U = \infty$, $D = 0$ of this last result: for it states that $\| L_j(\mathbf{x}) - \alpha_j \| < \epsilon \, (1 \leqslant j \leqslant n)$ is soluble for all $\boldsymbol{\alpha}$ by integral $\mathbf{x}$ provided that there are no integral $\mathbf{u} \neq \mathbf{0}$ such that $\| M_i(\mathbf{u}) \| = 0 \, (1 \leqslant i \leqslant m)$. In § 8 we shall prove a quantitative generalization of the general Kronecker theorem and in § 9 we shall indicate briefly another approach to transference theorems.

2. Transference between two homogeneous problems.
The theorems discussed in the introduction are easily deduced from:

THEOREM I. *Let $f_k(\mathbf{z})$ $(1 \leqslant k \leqslant l)$ be l linearly independent homogeneous linear forms in the l variables $\mathbf{z} = (z_1, \ldots, z_l)$ and let $g_k(\mathbf{w})$ be l linearly independent homogeneous linear forms in the l variables $\mathbf{w} = (w_1, \ldots, w_l)$ of determinant d. Suppose that all the products $z_i w_j$ $(1 \leqslant i, j \leqslant l)$ have integer coefficients in*

$$\Phi(\mathbf{z}, \mathbf{w}) = \sum_k f_k(\mathbf{z}) \, g_k(\mathbf{w}). \tag{1}$$

If the inequalities

$$|f_k(\mathbf{z})| \leqslant \lambda \quad (1 \leqslant k \leqslant l) \tag{2}$$

are soluble with integral $\mathbf{z} \neq \mathbf{0}$ *then the inequalities*

$$|g_k(\mathbf{w})| \leqslant (l-1)|\lambda d|^{1/(l-1)}, \tag{3}$$

are soluble with integral $\mathbf{w} \neq \mathbf{0}$.

Proof. By the hypothesis of linear independence the only solution of $f_k(\mathbf{z}) = 0$ $(1 \leqslant k \leqslant l)$ is $\mathbf{z} = \mathbf{0}$. Hence by the hypothesis of the theorem there is an integral $\mathbf{z} \neq \mathbf{0}$ such that

$$0 < \max |f_k(\mathbf{z})| \leqslant \lambda. \tag{4}$$

Since the right-hand side of (3) decreases when λ decreases we may suppose, by interchanging $f_1, \ldots, f_l$ if necessary, that

$$\max |f_k(\mathbf{z})| = f_l(\mathbf{z}) = \lambda > 0. \tag{5}$$

In what follows $\mathbf{z}$ is a fixed integral vector for which (5) holds.

Consider now the l linear forms

$$\Phi(\mathbf{z}, \mathbf{w}),$$

$$g_k(\mathbf{w}) \quad (k \neq l),$$

in the l variables $\mathbf{w}$. Their determinant is readily verified to be

$$f_l(\mathbf{z})\, d = \lambda d.$$

By Minkowski's linear forms theorem (Theorem III of Appendix B) there is thus an integral $\mathbf{w} \neq \mathbf{0}$ such that

$$\left.\begin{array}{l} |\Phi(\mathbf{z}, \mathbf{w})| < 1, \\ |g_k(\mathbf{w})| \leqslant |\lambda d|^{1/(l-1)} \quad (k \neq l). \end{array}\right\} \tag{6}$$

But $\Phi(\mathbf{z}, \mathbf{w})$ is an integer, by hypothesis, and so

$$\Phi(\mathbf{z}, \mathbf{w}) = 0.$$

Hence and by (5),

$$\lambda g_l(\mathbf{w}) = f_l(\mathbf{z})\, g_l(\mathbf{w}) = -\sum_{k \neq l} f_k(\mathbf{z})\, g_k(\mathbf{w}),$$

and so, by (5), (6),

$$|g_l(\mathbf{w})| \leqslant (l-1)|\lambda d|^{1/(l-1)}. \tag{7}$$

Since (3) follows at once from (6) and (7) this proves the theorem.

An almost immediate application gives

THEOREM II. *Let*

$$L_j(\mathbf{x}) = \sum_i \theta_{ji} x_i, \quad M_i(\mathbf{u}) = \sum_j \theta_{ji} u_j,$$

where $1 \leqslant i \leqslant m$, $1 \leqslant j \leqslant n$. *Suppose that there are integers* $\mathbf{x} \neq \mathbf{0}$ *such that*

$$\|L_j(\mathbf{x})\| \leqslant C, \quad |x_i| \leqslant X,$$

for some constants C and X, where $0 < C < 1 \leqslant X$. Then there are integers $\mathbf{u} \neq 0$ such that

$$\| M_i(\mathbf{u}) \| \leqslant D, \quad | u_j | \leqslant U, \tag{8}$$

where

$$D = (l-1) X^{(1-n)/(l-1)} C^{n/(l-1)}, \quad U = (l-1) X^{m/(l-1)} C^{(1-m)/(l-1)}, \tag{9}$$

and

$$l = m + n.$$

Proof. We introduce new variables

$$\mathbf{y} = (y_1, \ldots, y_n), \quad \mathbf{v} = (v_1, \ldots, v_m).$$

Put
$$f_k(\mathbf{x}, \mathbf{y}) = \begin{cases} C^{-1}(L_k(\mathbf{x}) + y_k) & (1 \leqslant k \leqslant n), \\ X^{-1} x_{k-n} & (n < k \leqslant l), \end{cases}$$

and
$$g_k(\mathbf{u}, \mathbf{v}) = \begin{cases} C u_k & (1 \leqslant k \leqslant n), \\ X(-M_{k-n}(\mathbf{u}) + v_{k-n}) & (n < k \leqslant l). \end{cases}$$

Then the f_k are linearly independent forms in the l variables $\mathbf{z} = (\mathbf{x}, \mathbf{y})$ and the g_k are linearly independent forms in the l variables $\mathbf{w} = (\mathbf{u}, \mathbf{v})$ with determinant

$$d = C^n X^m.$$

Further,
$$\sum_{k \leqslant l} f_k g_k = \sum_{j \leqslant n} u_j y_j + \sum_{i \leqslant m} v_i x_i,$$

since the terms in $x_i u_j$ all cancel out. By hypothesis there are integers $\mathbf{x}, \mathbf{y}$ such that

$$| f_k(\mathbf{x}, \mathbf{y}) | \leqslant 1,$$

so we may apply Theorem I with $\lambda = 1$. Hence there are integers $(\mathbf{u}, \mathbf{v}) \neq (\mathbf{0}, \mathbf{0})$ such that

$$\left. \begin{array}{c} C \, | u_j | \\ X \, | - M_i(\mathbf{u}) + v_i | \end{array} \right\} \leqslant (l-1)(C^n X^m)^{1/(l-1)} = \begin{cases} CU, \\ XD. \end{cases}$$

If $D < 1$, $\mathbf{u} = \mathbf{0}$ we should have $v_j = 0$; and so $(\mathbf{u}, \mathbf{v}) = \mathbf{0}$, which is excluded. Hence $\mathbf{u} \neq \mathbf{0}$, as asserted, or $D \geqslant 1$. But $U \geqslant 1$ since $X \geqslant 1 > C$ and so (8) is trivially soluble when $D \geqslant 1$.

COROLLARY. *The necessary and sufficient condition that there exist a constant $\gamma > 0$ such that*

$$(\max \| L_j(\mathbf{x}) \|)^n (\max | x_i |)^m \geqslant \gamma \tag{10}$$

for all integral $\mathbf{x} \neq \mathbf{0}$, *is that there exist a* $\delta > 0$ *such that*

$$(\max \| M_i(\mathbf{u}) \|)^m (\max | u_j |)^n \geq \delta \qquad (11)$$

for all integral $\mathbf{u} \neq \mathbf{0}$.

Proof. Let $\mathbf{x} \neq \mathbf{0}$ be integral and let

$$X = \max | x_i |, \quad C \geq \max \| L_j(\mathbf{x}) \| \quad (1 > C > 0). \qquad (12)$$

If $\delta > 0$ exists, then $\qquad D^m U^n \geq \delta$

for the D, U of (8), (9). But (9), (12) give

$$X^m C^n \geq (l-1)^{-l(l-1)} \delta^{l-1} = \gamma \quad \text{(say)}.$$

Similarly, the symmetry of the relation between the $L_j(\mathbf{x})$ and the $M_i(\mathbf{u})$ shows that if γ exists so does δ.

3. Application to simultaneous approximations.†

We prove a complement to Theorem VII of Chapter I (cf. Theorem VIII of Chapter I).

THEOREM III. *Let* $\theta_1, \ldots, \theta_n$ *be any n numbers in a real algebraic field of degree $n+1$ such that $1, \theta_1, \ldots, \theta_n$ are linearly independent over the rationals. Then there is a constant $\gamma > 0$ (depending only on $\theta_1, \ldots, \theta_n$) such that*

$$\max \| \theta_j x \| \geq \gamma x^{-1/n} \qquad (1)$$

for all integers $x > 0$.

Proof. By Theorem II, Corollary, it is enough to show the existence of a $\delta > 0$ such that

$$\| u_1 \theta_1 + \ldots + u_n \theta_n \| \geq \delta (\max | u_j |)^{-n} \qquad (2)$$

for all integral $\mathbf{u} \neq \mathbf{0}$. The left-hand side of (2) is

$$| v + u_1 \theta_1 + \ldots + u_n \theta_n | \leq \tfrac{1}{2} \qquad (3)$$

for some integer v. There is some rational integer $q \neq 0$ such that $q\theta_1, \ldots, q\theta_n$ are algebraic integers, and so

$$\alpha = qv + qu_1 \theta_1 + \ldots + qu_n \theta_n$$

is an algebraic integer, not 0 by hypothesis. Any of its n other algebraic conjugates

$$\alpha' = qv + qu_1 \theta_1' + \ldots + qu_n \theta_n',$$

† This section assumes a nodding acquaintance with algebraic number theory. It may be omitted at first reading.

satisfies the estimate

$$|\alpha'| \leqslant |\alpha| + |\alpha' - \alpha|$$
$$\leqslant \tfrac{1}{2}|q| + |qu_1(\theta_1' - \theta_1) + \dots + qu_n(\theta_n' - \theta_n)|$$
$$\leqslant E \max |u_j|$$

by (3), where E is independent of $\mathbf{u}$. On the other hand, the product of α with its n other conjugates is a rational non-zero integer, and so $\geqslant 1$ in absolute value. Hence

$$|\alpha| (E \max |u_j|)^n \geqslant 1,$$

which is (2) with $\delta = q^{-1} E^{-n}$.

As another application of Theorem II the reader will have no trouble in proving for himself

THEOREM IV. (Khintchine's transference principle.) *Let* $\theta_1, \dots, \theta_n$ *be any irrational numbers and let* $\omega_1 \geqslant 0, \omega_2 \geqslant 0$ *be the respective upper bounds of the numbers* ω, ω' *such that*

$$\|u_1\theta_1 + \dots + u_n\theta_n\| \leqslant (\max |u_j|)^{-n-\omega},$$
$$\max \|x\theta_j\| \leqslant x^{-(1+\omega')/n}$$

have infinitely many integer solutions. Then

$$\omega_1 \geqslant \omega_2 \geqslant \frac{\omega_1}{n^2 + (n-1)\omega_1},$$

with the obvious interpretation if ω_1 *or* ω_2 *is* ∞.

4. Transference between homogeneous and inhomogeneous problems. The key result here is

THEOREM V. *Let* $f_k(\mathbf{z})$ $(1 \leqslant k \leqslant l)$ *be* l *homogeneous linear forms in the* l *variables* $\mathbf{z} = (z_1, \dots, z_l)$ *of determinant* $\Delta \neq 0$; *and suppose that the only integer solution of*

$$\max |f_k(\mathbf{z})| < 1 \qquad (1)$$

is $\mathbf{z} = 0$. *Then for all real numbers* $\boldsymbol{\beta} = (\beta_1, \dots, \beta_l)$ *there are integer solutions of*

$$\max |f_k(\mathbf{z}) - \beta_k| < \tfrac{1}{2}(h+1), \qquad (2)$$

where

$$h = [|\Delta|]. \qquad (3)$$

Note 1. $|\Delta| \geqslant 1$ by Minkowski's linear forms theorem (Theorem III of Appendix B).

Note 2. That the theorem would be false if the right-hand side of (2) were replaced by $\frac{1}{2}|\Delta|$ is shown by the trivial example

$$f_1(\mathbf{z}) = \Delta z_1, \quad f_k(\mathbf{z}) = z_k \quad (k \neq 1),$$
$$\boldsymbol{\beta} = (\tfrac{1}{2}\Delta, 0, \ldots, 0) \quad (\Delta > 1).$$

Proof. Since $\Delta \neq 0$ there is always a $\boldsymbol{\zeta} = (\zeta_1, \ldots, \zeta_l)$, not in general integral, such that

$$f_k(\boldsymbol{\zeta}) = \beta_k \quad (1 \leqslant k \leqslant l).$$

It is convenient to define

$$F(\mathbf{z}) = \max |f_k(\mathbf{z})|, \tag{4}$$

so that $\quad\quad\quad F(\lambda \mathbf{z}) = |\lambda| F(\mathbf{z}) \tag{5}$

for numbers λ, and

$$F(\mathbf{z}^{(1)} + \mathbf{z}^{(2)}) \leqslant F(\mathbf{z}^{(1)}) + F(\mathbf{z}^{(2)}) \tag{6}$$

for any two vectors† $\mathbf{z}^{(1)}$ and $\mathbf{z}^{(2)}$. Hence (2) may be written

$$F(\mathbf{z} - \boldsymbol{\zeta}) < \tfrac{1}{2}(h+1).$$

For fixed $\boldsymbol{\zeta}$ there are clearly‡ only a finite number of integral $\mathbf{z}$ such that $F(\mathbf{z} - \boldsymbol{\zeta}) \leqslant F(\boldsymbol{\zeta})$ and, in particular, $F(\mathbf{z} - \boldsymbol{\zeta})$ attains its lower bound; say for $\mathbf{z}^{(0)}$. By writing $\mathbf{z} - \mathbf{z}^{(0)}$, $\boldsymbol{\zeta} - \mathbf{z}^{(0)}$ for $\mathbf{z}$, $\boldsymbol{\zeta}$ respectively, we may suppose, without loss of generality, that

$$F(\mathbf{z} - \boldsymbol{\zeta}) \geqslant F(\boldsymbol{\zeta}) \tag{7}$$

for all integral $\mathbf{z}$. We have to prove $F(\boldsymbol{\zeta}) < \tfrac{1}{2}(h+1)$.

We now introduce a new parameter u and consider the set of inequalities

$$F\left(\mathbf{z} - \frac{2u}{h+1}\boldsymbol{\zeta}\right) < 1, \tag{8}$$

$$|u| \leqslant |\Delta|. \tag{9}$$

in the $l+1$ variables $z_1, \ldots, z_l, u$. When F is replaced by its definition (4) we have $l+1$ homogeneous linear forms on the left-hand side with determinant Δ. Hence there are integers $z_1, \ldots, z_l, u$ not all 0 satisfying (8) and (9) by Theorem III of

† $F(\mathbf{z})$ is a convex distance function in the sense of Appendix B.
‡ Cf. Lemma 4 in Appendix B.

Appendix B. If $u = 0$ we should have an integral solution $\mathbf{z} \neq \mathbf{0}$ of (1) contrary to hypothesis. Hence, on taking $-\mathbf{z}$, $-u$ for $\mathbf{z}$, u if need be, we may suppose that

$$0 < u \leqslant h = [|\Delta|]. \tag{10}$$

But then, with this integral $\mathbf{z}$, we have

$$F(\mathbf{z} - \zeta) \leqslant F\left(\mathbf{z} - \frac{2u}{h+1}\zeta\right) + F\left(\frac{2u - h - 1}{h+1}\zeta\right)$$

$$< 1 + \left|\frac{2u - h - 1}{h+1}\right| F(\zeta)$$

$$\leqslant 1 + \frac{h-1}{h+1} F(\zeta),$$

by (6), (5), (10) respectively. Hence by (7) we have

$$F(\zeta) < 1 + \frac{h-1}{h+1} F(\zeta),$$

i.e.

$$F(\zeta) < \tfrac{1}{2}(h+1);$$

which is the assertion of the theorem.

As an immediate application we have

THEOREM VI. *Let $L_j(\mathbf{x})$, $\mathbf{x} = (x_1, \ldots, x_m)$ be n homogeneous forms in m variables. Suppose that there is no integer $\mathbf{x} \neq \mathbf{0}$ such that simultaneously*

$$\|L_j(\mathbf{x})\| < C, \quad |x_i| < X.$$

Then for all $\alpha_1, \ldots, \alpha_n$ the inequalities

$$\|L_j(\mathbf{x}) - \alpha_j\| \leqslant C_1, \quad |x_i| \leqslant X_1,$$

are soluble in integral $\mathbf{x}$, where

$$C_1 = \tfrac{1}{2}(h+1)C, \quad X_1 = \tfrac{1}{2}(h+1)X,$$

and

$$h = [X^{-m} C^{-n}].$$

Proof. This is an immediate application of Theorem V to the set of $l = m + n$ forms

$$f_k = \begin{cases} C^{-1}(L_k(\mathbf{x}) - y_k) & (1 \leqslant k \leqslant n) \\ X^{-1} x_{k-n} & (n < k \leqslant l) \end{cases}$$

of determinant $X^{-m} C^{-n}$ in the l variables

$$(x_1, \ldots, x_m, y_1, \ldots, y_n).$$

COROLLARY. *Suppose that*

$$C = \gamma X^{-m/n}$$

for some $\gamma > 0$. Then

$$X_1 = \tfrac{1}{2}([\gamma^{-n}] + 1)\, X, \quad C_1 = \tfrac{1}{2}([\gamma^{-n}] + 1)\, C,$$

so that
$$C_1 = \delta X_1^{-m/n},$$

where δ depends only on γ.

Proof. Clear.

The following simple theorem provides an indirect converse to Theorem VI, since it links the inhomogeneous problem for the L_j with the homogeneous problem for the M_i which is linked to the homogeneous problem for the L_j by Theorem II.

THEOREM VII. *Suppose that for all $\boldsymbol\alpha = (\alpha_1, \ldots, \alpha_n)$ there is an integer solution $\mathbf{x} = (x_1, \ldots, x_m)$ of*

$$\| L_j(\mathbf{x}) - \alpha_j \| < C_1, \quad |x_i| \leqslant X_1.$$

Then there is no integer solution $\mathbf{u} \neq \mathbf{0}$ of

$$\| M_i(\mathbf{u}) \| \leqslant D, \quad |u_j| \leqslant U,$$

where
$$D = (4mX_1)^{-1}; \quad U = (4nC_1)^{-1},$$

and L_j, M_i are transposed sets of forms as defined in § 1.

Proof. We use the identity

$$\sum_i x_i M_i(\mathbf{u}) = \sum_{i,j} \theta_{ji} x_i u_j = \sum_j u_j L_j(\mathbf{x}).$$

Suppose $\mathbf{u}$ exists. Choose any vector $\boldsymbol\alpha$ such that $\sum u_j \alpha_j = \tfrac{1}{2}$. Then

$$\begin{aligned}
\tfrac{1}{2} &= \| \sum u_j \alpha_j \| \\
&\leqslant \| \sum u_j(\alpha_j - L_j(\mathbf{x})) \| + \| \sum u_j L_j(\mathbf{x}) \| \\
&< nUC_1 + \| \sum x_i M_i(\mathbf{u}) \| \\
&\leqslant nUC_1 + mX_1 D \\
&= \tfrac{1}{4} + \tfrac{1}{4};
\end{aligned}$$

a contradiction.

COROLLARY. *If $C_1 = \gamma X_1^{-m/n}$ then $D = \delta U^{-n/m}$ where δ depends only on γ, m, n.*

Proof. Clear.

From the corollaries to Theorems II, VI, VII we have at once

THEOREM VIII. *The following four statements each imply all of the others.*

(i) *There is a constant $\gamma_1 > 0$ such that the inequalities*

$$\| L_j(\mathbf{x}) \| \leqslant \gamma_1 X^{-m/n}, \quad |x_i| \leqslant X,$$

are insoluble for all $X \geqslant 1$ in integers $\mathbf{x} \neq \mathbf{0}$.

(ii) *There is a constant $\gamma_2 > 0$ such that the inequalities*

$$\| M_i(\mathbf{u}) \| \leqslant \gamma_2 U^{-n/m}, \quad |u_j| \leqslant U,$$

are insoluble for all $U \geqslant 1$ in integers $\mathbf{u} \neq \mathbf{0}$.

(iii) *There is a constant $\gamma_3 > 0$ such that the inequalities*

$$\| L_j(\mathbf{x}) - \alpha_j \| \leqslant \gamma_3 X^{-m/n}, \quad |x_i| \leqslant X,$$

are soluble for all $X \geqslant 1$ and all $\boldsymbol{\alpha}$ in integral $\mathbf{x}$.

(iv) *There is a constant $\gamma_4 > 0$ such that the inequalities*

$$\| M_i(\mathbf{u}) - \beta_i \| \leqslant \gamma_4 U^{-n/m}, \quad |u_j| \leqslant U,$$

are soluble for all $U \geqslant 1$ and all $\boldsymbol{\beta}$ in integral $\mathbf{u}$.

5. A direct converse to Theorem V.† We need

LEMMA 1. *Let $\mathcal{R}$ be a closed, convex l-dimensional region symmetric about $\mathbf{0}$ and containing $(0, \ldots, 0, \pm \mu)$ for some $\mu > 0$. Let $\mathcal{R}_0$ be the $(l-1)$-dimensional set of points, $\mathbf{x} = (x_1, \ldots, x_{l-1})$ such that $(\mathbf{x}, y) \in \mathcal{R}$ for at least one y. Then*

$$lV \geqslant 2V_0 \mu,$$

where V, V_0 are the l, $(l-1)$-dimensional volumes of $\mathcal{R}$, $\mathcal{R}_0$ respectively.

Proof. For given $\mathbf{x} \in \mathcal{R}_0$ the set of y such that $(\mathbf{x}, y) \in \mathcal{R}$ is an interval, say $\eta_1(\mathbf{x}) \leqslant y \leqslant \eta_2(\mathbf{x})$. The region $\mathcal{S}$ of points $(\mathbf{x}, y)$ where

$$\mathbf{x} \in \mathcal{R}_0, \quad |y| \leqslant \tfrac{1}{2}(\eta_2(\mathbf{x}) - \eta_1(\mathbf{x})) = Y(\mathbf{x}),$$

has clearly volume V. If $\mathbf{x}^{(1)}, \mathbf{x}^{(2)} \in \mathcal{R}_0$ then by convexity $\mathcal{R}$ contains the whole of the plane quadrilateral with vertices

$$(\mathbf{x}^{(i)}, \eta_j(\mathbf{x}^{(i)})) \quad (i, j = 1, 2),$$

and so $\mathcal{S}$ contains the whole quadrilateral with vertices

$$(\mathbf{x}^{(i)}, \pm Y(\mathbf{x}^{(i)})) \quad (i = 1, 2).$$

† This section may be omitted at first reading.

In particular the line joining any two points $(\mathbf{x}^{(i)}, y^{(i)})$ of $\mathscr{S}$ lies in $\mathscr{S}$; i.e. $\mathscr{S}$ is convex.

By hypothesis $(0, \ldots, 0, \pm\mu) \in \mathscr{S}$ and by construction $(\mathbf{x}, 0) \in \mathscr{S}$ whenever $\mathbf{x} \in \mathscr{R}_0$. Hence by convexity $\mathscr{S}$ contains the 'double cone'

$$(\lambda\mathbf{x}, \pm(1-\lambda)\mu), \quad 0 \leqslant \lambda \leqslant 1, \mathbf{x} \in \mathscr{R}_0.$$

It is readily verified that the double cone has volume $2l^{-1}\mu V_0$. Since it is contained in $\mathscr{S}$ of volume V the lemma is established.

THEOREM IX. (Birch.) *Let* $f_k(\mathbf{z})$ $(1 \leqslant k \leqslant l)$ *be linear forms in*

$$\mathbf{z} = (z_1, \ldots, z_l)$$

of determinant $\Delta \neq 0$. *Suppose that to every* $\boldsymbol{\zeta}$ *there is an integral* $\mathbf{z}$ *with*

$$|f_k(\boldsymbol{\zeta}-\mathbf{z})| \leqslant 1 \quad (1 \leqslant k \leqslant l).$$

Then

$$\max|f_k(\mathbf{z})| \geqslant l^{-1} 2^{-l+1} |\Delta|$$

for all integral $\mathbf{z} \neq 0$.

Proof. Let $\mathbf{z}^{(0)} \neq 0$ be integral and let $\max|f_k(\mathbf{z}^{(0)})| = \lambda_0$. By Lemma 7 of Appendix B we may suppose without loss of generality that $\mathbf{z}^{(0)} = (0, 0, \ldots, 0, z_{0l})$. Since $|z_{0l}| \geqslant 1$ the points $(0, 0, \ldots, 0, \pm\lambda_0^{-1})$ satisfy $\max|f_k(\mathbf{z})| \leqslant 1$.

Let $\mathscr{R}$ be defined by $|f_k(\mathbf{z})| \leqslant 1$ $(1 \leqslant k \leqslant l)$ and let $\mathscr{R}_0, V, V_0$ be as in the proof of Lemma 1. To any $\boldsymbol{\zeta} = (\zeta_1, \ldots, \zeta_l)$ there is an $\mathbf{x} \equiv (\zeta_1, \ldots, \zeta_{l-1})$ modulo 1 in $\mathscr{R}_0$ by hypothesis, so $V_0 \geqslant 1$. The theorem now follows from Lemma 1 with

$$V = 2^l |\Delta|^{-1}, \quad \mu = \lambda_0^{-1}, \quad V_0 \geqslant 1.$$

6. Application to inhomogeneous approximation. In
§§ 6, 7 we shall use the techniques developed in §§ 1–4 to investigate how far the results of Chapter III are best possible. Our first goal is

THEOREM X. *For all pairs of integers* $m > 0$, $n > 0$ *there is a constant* $\Gamma_{m,n} > 0$ *with the following property. Let* $L_j(\mathbf{x})$ *be any* n *homogeneous forms in* m *variables. Then there is an* $\boldsymbol{\alpha} = (\alpha_1, \ldots, \alpha_n)$ *such that*

$$(\max\|L_j(\mathbf{x}) - \alpha_j\|)^m (\max|x_i|)^n \geqslant \Gamma_{m,n}$$

for all integers $\mathbf{x} \neq 0$.

THEOREM XI. *In particular, $\Gamma_{1,1}$ may be taken to be* $(51)^{-1}$.

What the best possible value of $\Gamma_{m,n}$ is remains unknown even for $m = n = 1$. Theorem XI asserts, in particular, that to every θ there is an α such that $|x| \, \|\theta x - \alpha\| \geqslant (51)^{-1}$ for all integers $x \neq 0$; and so is a counterpart to Theorem II of Chapter III.

The proof of Theorem X requires three lemmas relating to the transposed system of forms $M_i(\mathbf{u})$.

LEMMA 2. *Let* $\mathbf{u}^{(r)} = (u_{r1}, ..., u_{rn}) \neq \mathbf{0}$ *for* $r = 1, 2, ...$ *be a finite or infinite sequence of integer vectors. Define* $\rho_r > 0$ *by*

$$\rho_r^2 = u_{r1}^2 + ... + u_{rn}^2. \tag{1}$$

Suppose that $\qquad \rho_{r+1} \geqslant k\rho_r \quad (r = 1, 2, ...),$ $\tag{2}$

for some number $k > 2$. Then there is a set of real numbers

$$\boldsymbol{\alpha} = (\alpha_1, ..., \alpha_n),$$

such that

$$\|\mathbf{u}^{(r)}\boldsymbol{\alpha}\| = \|u_{r1}\alpha_1 + ... + u_{rn}\alpha_n\| \geqslant \frac{1}{2}\left(1 - \frac{1}{k-1}\right) \tag{3}$$

for all r.

Proof. The planes

$$\mathbf{u}^{(r)}\mathbf{z} = \text{integer}$$

in the space of $\mathbf{z} = (z_1, ..., z_n)$ lie ρ_r^{-1} apart (in the usual Euclidean metric). The perpendicular distance from a general $\mathbf{z}$ to the nearest of those planes is $\rho_r^{-1}\|\mathbf{u}^{(r)}\mathbf{z}\|$. We shall construct a sequence of spheres $\mathscr{C}_r$ each contained in its predecessors, with radius

$$1/2(k-1)\rho_r \tag{4}$$

and centre on a plane

$$\mathbf{u}^{(r)}\mathbf{z} = \text{integer} + \tfrac{1}{2}. \tag{5}$$

Then (3) will hold for all points of $\mathscr{C}_r$ from the geometric interpretation of $\mathbf{u}^{(r)}\mathbf{z}$. Since each sphere contains its successor there is a point $\boldsymbol{\alpha}$ contained in all of them, which clearly does what is required.

It remains to construct the $\mathscr{C}_r$. We take for $\mathscr{C}_1$ any sphere with the right radius and centre on $\mathbf{u}^{(1)}\mathbf{z} = \tfrac{1}{2}$. If $\mathscr{C}_{r-1}$ has already been constructed, with centre $\boldsymbol{\beta}_{r-1}$, say, there is a point $\boldsymbol{\beta}_r$ on one of the planes (5), at most a distance $\frac{1}{2}\rho_r^{-1}$ from β_{r-1} (e.g. the foot of the perpendicular from β_{r-1} on the nearest plane (5)). Take for $\mathscr{C}_r$ the sphere with centre β_r and radius (4). Then $\mathscr{C}_r$ lies in $\mathscr{C}_{r-1}$ since

$$\frac{1}{2\rho_r} + \frac{1}{2(k-1)\rho_r} \leqslant \frac{1}{2(k-1)\rho_{r-1}}$$

by (2). Hence we may construct $\mathscr{C}_1, \mathscr{C}_2, \ldots$ and the lemma is established.

For all numbers $\rho \geqslant 1$ it is convenient to define $\eta(\rho)$ to be the minimum of

$$\max_i \| M_i(\mathbf{u}) \|$$

taken over all integer vectors $\mathbf{u} \neq \mathbf{0}$ in

$$u_1^2 + \ldots + u_n^2 \leqslant \rho^2.$$

LEMMA 3. (i) $\eta(\rho)$ *does not increase if ρ increases.*
(ii) *There is a constant $\gamma_{m,n}$ depending only on m, n such that*

$$(\eta(\rho))^m \rho^n \leqslant \gamma_{m,n}. \tag{6}$$

(iii) *We may take $\gamma_{1,1} = 1$.*
Proof. (i) Trivial.
(ii), (iii) By Theorem VI of Chapter I we may satisfy

$$| u_j | \leqslant n^{-\frac{1}{2}}\rho, \quad \| M_i(\mathbf{u}) \| < (n^{-\frac{1}{2}}\rho)^{-n/m},$$

with an integer $\mathbf{u} \neq \mathbf{0}$ if $\rho > n^{\frac{1}{2}}$. The result then follows at once. If $\rho \leqslant n^{\frac{1}{2}}$ then it is trivial since $\eta(\rho) \leqslant \frac{1}{2}$.

[An alternative proof, which gives a better estimate for $\gamma_{m,n}$ (except $\gamma_{1,1}$), uses Theorem IV of Appendix B and the fact that the region $| M_i(\mathbf{u}) + v_i | \leqslant D$, $\Sigma u_j^2 \leqslant \rho^2$ in the space of

$$(u_1, \ldots, u_n, v_1, \ldots, v_m)$$

is convex for any $D > 0$, $\rho > 0$.]

LEMMA 4. *For any $k > 1$ we may find a sequence of integral vectors* $\mathbf{u}^{(r)} = (u_{r1}, \ldots, u_{rn}) \neq \mathbf{0}$ *for* $r = 1, 2, \ldots$, *with ρ_r defined by* (1), *such that*

$$\rho_1 \leqslant k, \tag{7}$$

$$\rho_{r+1} \geqslant k\rho_r \quad (r = 1, 2, \ldots), \tag{8}$$

$$\max_i \| M_i(\mathbf{u}^{(r)}) \| = \eta(k^{-1}\rho_{r+1}). \tag{9}$$

The sequence is infinite unless there is an integral $\mathbf{u} \neq \mathbf{0}$ *with* $\| M_i(\mathbf{u}) \| = 0$ $(1 \leqslant i \leqslant m)$. *If there is such a* $\mathbf{u}$ *the sequence terminates with a* $\mathbf{u}^{(R)}$ *such that* $\max \| M_i(\mathbf{u}^{(R)}) \| = 0$ *but* $\max \| M_i(\mathbf{u}^{(r)}) \| \neq 0$ *for* $r < R$.

Proof. Suppose first that there is an integral $\mathbf{u} \neq \mathbf{0}$ with $\| M_i(\mathbf{u}) \| = 0$ $(1 \leqslant i \leqslant m)$. We first construct a sequence of integral vectors $\mathbf{v}^{(r)} \neq \mathbf{0}$ and numbers $\sigma_r > 0$ with

$$\sigma_r^2 = v_{r1}^2 + \ldots + v_{rn}^2$$

according to the following prescription:

(i) $\mathbf{v}^{(1)} \neq \mathbf{0}$ is an integer vector with

$$\| M_i(\mathbf{v}^{(1)}) \| = 0 \quad (1 \leqslant i \leqslant m),$$

for which σ_1 is as small as possible.

(ii) If $\mathbf{v}^{(1)}, \ldots, \mathbf{v}^{(R)}$ have been constructed for some R and $\sigma_R \leqslant k$ then the sequence stops with $\mathbf{v}^{(R)}$.

(iii) If $\mathbf{v}^{(1)}, \ldots, \mathbf{v}^{(r)}$ have been constructed and $\sigma_r > k$ then $\mathbf{v}^{(r+1)} \neq \mathbf{0}$ is an integer vector with

$$\sigma_{r+1} \leqslant k^{-1}\sigma_r; \quad \max \| M_i(\mathbf{v}^{(r+1)}) \| = \eta(k^{-1}\sigma_r),$$

which exists by the definition of $\eta(\rho)$.

Since $\sigma_{r+1} \leqslant k^{-1}\sigma_r$ the sequence does terminate with a $\mathbf{v}^{(R)}$. Then the $\mathbf{u}^{(r)} = \mathbf{v}^{(R+1-r)}$ clearly do what is required.

If there is no integral $\mathbf{u} \neq \mathbf{0}$ with $\| M_i(\mathbf{u}) \| = 0$ $(1 \leqslant i \leqslant m)$ the proof is more roundabout. Let $\epsilon > 0$ be arbitrarily small and let

$$\mathbf{v}^{(1, \, \epsilon)}, \ldots, \mathbf{v}^{(R, \, \epsilon)},$$

where $R = R(\epsilon)$ depends on ϵ, be any sequence of vectors constructed by (ii), (iii) above and

(i′) $\mathbf{v}^{(1, \, \epsilon)} \neq \mathbf{0}$ is any integral vector with

$$\max \| M_i(\mathbf{v}^{(1, \, \epsilon)}) \| < \epsilon.$$

The $\mathbf{u}^{(r, \, \epsilon)} = \mathbf{v}^{(R+1-r, \, \epsilon)}$ satisfy (7) and also (8), (9) for $r < R$.

By (7) there are only a finite number of possibilities for $\mathbf{u}^{(1, \, \epsilon)}$, so one of them, say $\mathbf{u}^{(1)}$, must occur for arbitrarily small† ϵ. Since

---

† I.e. for any $\epsilon_0 > 0$ there is an ϵ with $0 < \epsilon < \epsilon_0$ and $\mathbf{u}^{(1, \, \epsilon)} = \mathbf{u}^{(1)}$.

$\max \| M_i(\mathbf{u}^{(1)}) \| \neq 0$, by hypothesis, we must have $R(\epsilon) \geqslant 2$ for small ϵ by (i'). Since $\eta(\rho) \to 0$ as $\rho \to \infty$ by Lemma 3, there are at most a finite number of vectors $\mathbf{u}^{(2, \epsilon)}$ which satisfy (9) with $r = 1$ and the chosen $\mathbf{u}^{(1)} = \mathbf{u}^{(1, \epsilon)}$. We choose $\mathbf{u}^{(2)}$ such that $\mathbf{u}^{(1, \epsilon)} = \mathbf{u}^{(1)}$, $\mathbf{u}^{(2, \epsilon)} = \mathbf{u}^{(2)}$ occur together for arbitrarily small ϵ. Suppose that $\mathbf{u}^{(1)}, \dots, \mathbf{u}^{(r)}$ have already been chosen so that

$$\mathbf{u}^{(1, \epsilon)} = \mathbf{u}^{(1)}, \dots, \mathbf{u}^{(r, \epsilon)} = \mathbf{u}^{(r)} \tag{10}$$

occur simultaneously for arbitrarily small ϵ. Since

$$\max \| M_i(\mathbf{u}^{(r)}) \| \neq 0,$$

by hypothesis, we must have $R(\epsilon) \geqslant r + 1$ by (i') if ϵ is small enough and (10) holds. As before, (9) shows that there are only a finite number of candidates for $\mathbf{u}^{(r+1, \epsilon)}$ compatible with (10); so one of them, say $\mathbf{u}^{(r+1)}$, must occur for arbitrarily small ϵ. The $\mathbf{u}^{(1)}, \mathbf{u}^{(2)} \dots$ constructed in this way clearly have all the required properties.

Proof of Theorem X. Let $\mathbf{u}^{(r)}$ be the vectors constructed in Lemma 4 for

$$k = 3,$$

and let $\boldsymbol{\alpha}$ be then constructed by Lemma 2, so that

$$\| \mathbf{u}^{(r)} \boldsymbol{\alpha} \| \geqslant \tfrac{1}{4} \tag{11}$$

for $1 \leqslant r \leqslant R$ or $1 \leqslant r < \infty$ as the case may be. Let $\mathbf{x} \neq \mathbf{0}$ be integral and put

$$\max_j \| L_j(\mathbf{x}) - \alpha_j \| = C, \quad \max_i | x_i | = X.$$

As in the proof of Theorem VII we have

$$\| \mathbf{u}^{(r)} \boldsymbol{\alpha} \| \leqslant n \rho_r C + m X D_r \tag{12}$$

with, by (9),

$$D_r = \max \| M_i(\mathbf{u}^{(r)}) \| = \begin{cases} \eta(\tfrac{1}{3}\rho_{r+1}) & (r \neq R), \\ 0 & (r = R), \end{cases} \tag{13}$$

since $\max \| u_{rj} \| \leqslant \rho_r$.

Suppose, first, that we can choose an integer r such that

$$mD_{r-1}X \geqslant \tfrac{1}{8} \geqslant mD_r X. \tag{14}$$

Then by (11), (12) we have

$$n\rho_r C \geqslant \tfrac{1}{8},$$

and so
$$X^m C^n \geqslant \frac{1}{(8m)^m (8n)^n D_{r-1}^m \rho_r^n} \geqslant \Gamma'_{m, n} \qquad (15)$$

for some $\Gamma'_{m, n} > 0$ by Lemma 3 and (13).

Such an r exists unless

$$mD_1 X < \tfrac{1}{8},$$

and then
$$n\rho_1 C > \tfrac{1}{8}$$

by (11), (12). Since $\rho_1 \leqslant k = 3$ by (7) and trivially

$$X = \max |x_i| \geqslant 1,$$

we have
$$X^m C^n \geqslant C^n \geqslant (8n\rho_1)^{-n} \geqslant \Gamma''_{m, n},$$

for some $\Gamma''_{m, n} > 0$. This completes the proof of Theorem X on putting $\Gamma_{m, n} = \min (\Gamma'_{m, n}, \Gamma''_{m, n})$.

COROLLARY. *Suppose that $\rho^n(\eta(\rho))^m \to 0$ as $\rho \to \infty$. Then*

$$(\max \| L_j(\mathbf{x}) - \alpha_j \|)^n (\max |x_i|)^m \leqslant M,$$

with the $\boldsymbol{\alpha}$ just constructed, has only a finite number of solutions for any M, however large.

Proof. Suppose first that there is some $\mathbf{u}^{(R)}$ with $D_R = 0$. Then (11), (12) shows that
$$C \geqslant (4n\rho_R)^{-1}.$$

Since there are only a finite number of $\mathbf{x}$ with $\max |x_i|$ below any given bound, the corollary is true in this case.

Otherwise r takes all positive values and the central expression of (15) tends to ∞ as $r \to \infty$ since $D_{r-1} = \eta(\tfrac{1}{3}\rho_r)$. But there are only a finite number of solutions $\mathbf{x}$ of (14) for each r and again the corollary is true.

To obtain the estimate $(51)^{-1}$ for $\Gamma_{1,1}$ we must refine the preceding argument.

LEMMA 5. *Let λ, μ, ν be non-negative and $\kappa > 1$. If*

$$\lambda\mu \leqslant \nu^2, \quad \lambda \leqslant \kappa\nu, \quad \mu \leqslant \kappa\nu,$$

then
$$\lambda + \mu \leqslant (\kappa + \kappa^{-1}) \nu.$$

Proof. If $\lambda \leqslant \nu$, $\mu \leqslant \nu$ there is nothing to prove since $\kappa + \kappa^{-1} > 2$. If, say, $\nu < \lambda = \xi \nu$ then $1 < \xi \leqslant \kappa$ and $\mu \leqslant \lambda^{-1} \nu^2 \leqslant \xi^{-1} \nu$. Hence $\lambda + \mu \leqslant (\xi + \xi^{-1}) \nu \leqslant (\kappa + \kappa^{-1}) \nu$.

Proof of Theorem XI. For $m = n = 1$ we may simplify the notation by writing $x, u, u^{(r)}, \alpha$ for $x_1, u_1, u_{r1}, \alpha_1$ respectively, so that $\rho_r = |u^{(r)}|$. Further, $L_1(\mathbf{x}) = \theta x$, $M_1(\mathbf{u}) = \theta u$ for some number θ.

Let $k > 2$ be a number to be fixed later and let $u^{(r)}$, α be determined by Lemmas 4 and 2, so that

$$\| u^{(r)} \alpha \| \geqslant \frac{1}{2} \left(1 - \frac{1}{k-1} \right). \tag{16}$$

For any integer $x \neq 0$ we have, as before,

$$\| u^{(r)} \alpha \| = \| u^{(r)}(\alpha - \theta x) + x u^{(r)} \theta \|$$
$$\leqslant | u^{(r)} | \, \| \theta x - \alpha \| + | x | \, \| u^{(r)} \theta \|. \tag{17}$$

Suppose, first, that there is some integer r such that

$$| x | \, \| u^{(r)} \theta \| \geqslant (k | x | \, \| \theta x - \alpha \|)^{\frac{1}{2}} \geqslant | x | \, \| u^{(r+1)} \theta \|. \tag{18}$$

Now
$$| u^{(r+1)} | \, \| u^{(r)} \theta \| \leqslant k \tag{19}$$

by Lemma 3 (iii) and (9); so the left-hand part of (18) implies

$$| u^{(r+1)} | \, \| \theta x - \alpha \| \leqslant (k | x | \, \| \theta x - \alpha \|)^{\frac{1}{2}}. \tag{20}$$

But $| u^{(r+1)} | \, \| u^{(r+1)} \theta \| \leqslant 1$ by (8) and (19); so Lemma 5 applies with
$$\lambda = | x | \, \| u^{(r+1)} \theta \|, \quad \mu = | u^{(r+1)} | \, \| \theta x - \alpha \|,$$
$$\nu^2 = | x | \, \| \theta x - \alpha \|, \quad \kappa = k^{\frac{1}{2}},$$

by (20) and the right-hand part of (18). Hence the right-hand side of (17) is

$$\leqslant (k^{\frac{1}{2}} + k^{-\frac{1}{2}})(| x | \, \| \theta x - \alpha \|)^{\frac{1}{2}}. \tag{21}$$

From (16), (17), (21) we have

$$| x | \, \| \theta x - \alpha \| \geqslant (k-2)^2 k / 4(k^2 - 1)^2.$$

This expression has its maximum near $k = 11/2$ and on putting $k = 11/2$ it becomes
$$\frac{539}{27,378} > \frac{1}{51}.$$

There is an integer r satisfying (18) except when

$$(k\,|\,x\,|\,\|\,\theta x-a\,\|)^{\frac{1}{2}} > |\,x\,|\,\|\,u^{(1)}\theta\,\|.$$

If also $(k\,|\,x\,|\,\|\,\theta x-\alpha\,\|)^{\frac{1}{2}} \geqslant |\,u^{(1)}\,|\,\|\,\theta x-\alpha\,\|,$

the preceding argument continues to apply with $r+1=1$. Otherwise, since $|\,u^{(1)}\,| \leqslant k$ and $|\,x\,| \geqslant 1$, we have

$$(k\,|\,x\,|\,\|\,\theta x-\alpha\,\|)^{\frac{1}{2}} \leqslant |\,u^{(1)}\,|\,\|\,\theta x-\alpha\,\|$$
$$\leqslant k\,|\,x\,|\,\|\,\theta x-\alpha\,\|,$$

and $|\,x\,|\,\|\,\theta x-\alpha\,\| \geqslant k^{-1}=2/11 > 1/51.$

7. Regular and singular systems.

We shall say that the system $L_j(\mathbf{x})$ of n forms in m variables is SINGULAR if, for each $\epsilon > 0$, the set of inequalities

$$\|\,L_j(\mathbf{x})\,\| < \epsilon X^{-m/n}, \quad |\,x_i\,| \leqslant X \tag{1}$$

has an integer solution $\mathbf{x} \neq 0$ for all X greater than some $X_0(\epsilon)$. Otherwise the system is REGULAR.

[The justification of the nomenclature is that the coefficients θ_{ji} of systems of singular forms are a set of measure 0 in the space of mn dimensions, as we now prove. For integer $\mathbf{x}$ the value of $\|\,L_j(\mathbf{x})\,\|$ depends only on the θ_{ji} modulo 1, so we may confine ourselves to $0 \leqslant \theta_{ji} < 1.$

For fixed integral $\mathbf{x} \neq 0$ with, say, $x_1 \neq 0$ and for fixed $\theta_{j2}, ..., \theta_{jm}$ the inequality $\|\,L_j(\mathbf{x})\,\| \leqslant \epsilon X^{-m/n}$ restricts θ_{j1} to a set of measure $2\epsilon X^{-m/n}$; and so for fixed $\mathbf{x} \neq 0$ the set of θ_{ji} with

$$\|\,L_j(\mathbf{x})\,\| \leqslant \epsilon X^{-m/n} \quad (1 \leqslant j \leqslant n)$$

has measure $(2\epsilon)^n X^{-m}$. But there are $(2X+1)^m - 1 < (3X)^m$ integral $\mathbf{x} \neq 0$ with $\max |\,x_i\,| \leqslant X$, and so (1) is soluble with fixed X for a set of θ_{ji} of measure at most $\epsilon_1 = 3^m 2^n \epsilon^n$. Hence, by the Borel-Cantelli lemma,† the set of θ_{ji} such that (1) is soluble for all X greater than some X_0 depending on the θ_{ji} has measure at most ϵ_1. A fortiori the set of singular θ_{ji} has measure at most ϵ_1, and so measure 0 since ϵ is arbitrarily small.]

---

† Namely that $|\cup_{r \geqslant R} \cap \mathscr{E}_r| \leqslant \lim\inf |\mathscr{E}_r|$ for any sequence of sets $\mathscr{E}_r$ $(r \geqslant 1)$, where $\cup, \cap$ denote the union and intersection respectively and $|\mathscr{G}|$ is the measure of $\mathscr{G}$. For $|\cap_{r \geqslant R} \mathscr{E}_r| \leqslant |\mathscr{E}_R|$, and $|\cup_R \mathscr{F}_R| = \lim_R |\mathscr{F}_R|$ for any sequence $\mathscr{F}_R$ such that $\mathscr{F}_R$ contains $\mathscr{F}_S$ whenever $R \leqslant S$.

THEOREM XII. *The necessary and sufficient condition that the system $L_j(\mathbf{x})$ be singular is that the transposed system $M_i(\mathbf{u})$ be singular.*

This follows from Theorem II by a similar argument to that used in proving its corollary, so we suppress the proof. The following result may be regarded as a partial generalization of Theorem II of Chapter III.

THEOREM XIII. *The necessary and sufficient condition that the system $L_j(\mathbf{x})$ be regular is that there is a number $\delta > 0$ such that*

$$(\max_j \| L_j(\mathbf{x}) - \alpha_j \|)^n (\max_i | x_i |)^m < \delta, \tag{2}$$

has infinitely many integral solutions $\mathbf{x}$ for each real $\boldsymbol{\alpha}$.

Proof. Suppose first that the system $L_j(\mathbf{x})$ is regular; that is, there is some $\gamma > 0$ such that the inequalities

$$\| L_j(\mathbf{x}) \| \leqslant \gamma X^{-m/n}, \quad | x_i | \leqslant X, \tag{3}$$

are insoluble for some arbitrarily large value of X. By Theorem VI, Corollary, there is a solution $\mathbf{x}$ of

$$\| L_j(\mathbf{x}) - \alpha_j \| \leqslant \delta_1 X_1^{-m/n}, \quad | x_i | \leqslant X_1 = \lambda X, \tag{4}$$

for each $\boldsymbol{\alpha}$, where δ_1, λ depend only on γ: and so (2) holds for $\delta = \delta_1^n$. As $X \to \infty$ through the values for which (3) is insoluble we must get infinitely many solutions of (2) in this way; except possibly when there is an integral $\mathbf{x}^{(0)}$ such that

$$\| L_j(\mathbf{x}^{(0)}) - \alpha_j \| = 0 \quad (1 \leqslant j \leqslant n).$$

But, by Theorem VI of Chapter I, there are certainly integer solutions $\mathbf{x} \neq \mathbf{0}$ of

$$\| L_j(\mathbf{x}) \| \leqslant X^{-m/n}, \quad | x_i | \leqslant X, \tag{5}$$

for all X. Since $L_j(\mathbf{x})$ is regular we obtain infinitely many $\mathbf{x}$ as $X \to \infty$ by the definition of regularity. On writing $\mathbf{x} - \mathbf{x}^{(0)}$ for $\mathbf{x}$ in (5) we have a solution $\mathbf{x} \neq \mathbf{x}^{(0)}$ of

$$\| L_j(\mathbf{x}) - \alpha_j \| \leqslant X^{-m/n}, \quad | x_i | \leqslant X + X_0,$$

where $X_0 = \max (| x_{01} |, ..., | x_{0m} |)$. Since X_0 is fixed this gives infinitely many solutions of (2) for any $\delta > 1$ on letting $X \to \infty$.

If, however, the system is singular then the function $\eta(\rho)$ defined on page 87 in § 6 satisfies

$$\rho^n (\eta(\rho))^m \to 0 \quad (\rho \to \infty),$$

by Theorem XII and since

$$\max |u_j|^2 \leqslant \rho^2 = u_1^2 + \ldots + u_n^2 \leqslant n \max |u_j|^2.$$

The truth of our theorem then follows at once from Theorem X, Corollary.

When $m = n = 1$ so that

$$L_1(\mathbf{x}) = \theta x, \quad \theta = \theta_{11}, \quad x = x_1,$$

it is readily seen that the singular systems are just those in which θ is rational.† For if $x = p/q$ where p, q are integers, then $x = q$ is a solution of (1) for any $\epsilon > 0$ as soon as $X > q$. On the other hand if θ is irrational and p_n/q_n are the successive best approximations (in the language of Chapter I) we have no solution of

$$\|x\theta\| < \|q_n\theta\|, \quad 0 < x < q_{n+1}$$

for any n and, by (16) of Chapter I, $q_{n+1}\|q_n\theta\| > \frac{1}{2}$. However, except for $m = n = 1$, there exist non-trivial singular systems. By Theorem XII in proving this we may suppose without loss of generality that $m \geqslant n$ so $m \geqslant 2$. We confine ourselves to the simplest but quite typical case $n = 1$, $m = 2$. For a later application we prove then something much stronger than mere existence.

THEOREM XIV. *Let* $\omega(t) > 0$ *for* $t = 1, 2, \ldots$. *Then there are numbers* θ, ϕ *such that*
   (A) *The pair of inequalities*

$$\|r\theta + s\phi\| < \omega(t), \quad 0 < \max(|r|, |s|) \leqslant t,$$

is soluble in integers r, s *for all* $t = 1, 2, \ldots$.
   (B) $\|r\theta + s\phi\| \neq 0$ *for all integers* $(r, s) \neq (0, 0)$.
   Note. We shall be interested in the case when $\omega(t) \to 0$ rapidly as $t \to \infty$. If $t^2\omega(t) \to 0$ the system is singular by definition.
   Proof. We may suppose without loss of generality that $\omega(t)$ decreases monotonely to 0, by using $\min(t^{-1}, \omega(1), \omega(2), \ldots, \omega(t))$ instead of $\omega(t)$ if need be.

† The truth of this statement also follows by comparing Theorem II of Chapter III and Theorem XIII.

It is convenient to use a geometric representation and to take θ, ϕ as rectangular co-ordinates. We shall find a sequence of integers

$$1 = t_1 < t_2 < t_3 < \ldots,$$

and a sequence of squares:

$$\mathscr{S}_j : \ |\theta - \theta_j| \leqslant \delta_j, \quad |\phi - \phi_j| \leqslant \delta_j.$$

The squares $\mathscr{S}_j$ will satisfy four conditions:

(i)$_j$ if $t \leqslant t_j$ there are integers r, s such that

$$\|r\theta + s\phi\| < \omega(t), \quad 0 < \max(|r|, |s|) \leqslant t$$

for all $(\theta, \phi) \in \mathscr{S}_j$.

(ii)$_j$ if $0 < \max(|r|, |s|) < t_j$ then $\|r\theta + s\phi\| \neq 0$ for all $(\theta, \phi) \in \mathscr{S}_j$.

(iii)$_j$ the centre (θ_j, ϕ_j) of $\mathscr{S}_j$ lies on some line

$$r_j \theta_j + s_j \phi_j = l_j, \quad t_j = \max(|r_j|, |s_j|),$$

where r_j, s_j, l_j are integers.

(iv)$_j$ $\mathscr{S}_j$ is contained in $\mathscr{S}_{j-1}$ $(j > 1)$.

We remark first that if we can find the $\mathscr{S}_j$ the lemma is proved. By (iv)$_j$ there must be a point $(\theta_\infty, \phi_\infty)$ which lies in all $\mathscr{S}_j$. Then $(\theta_\infty, \phi_\infty)$ have the required properties, by (i)$_j$ and (ii)$_j$.

We take for $\mathscr{S}_1$ the square

$$|\theta - \theta_1| \leqslant \tfrac{1}{3}\omega(1), \quad |\phi| \leqslant \tfrac{1}{3}\omega(1)$$

for any θ_1. Then (i)$_1$, (iii)$_1$ are satisfied with $s_1 = t_1 = 1$, $r_1 = l_1 = 0$ and (ii)$_1$, (iv)$_1$ are vacuous. We now suppose that $t_1, \ldots, t_j$, $\mathscr{S}_1, \ldots, \mathscr{S}_j$ have been constructed and will construct t_{j+1}, $\mathscr{S}_{j+1}$. There are certainly infinitely many distinct lines $r\theta + s\phi = l$ $(r, s, l$ integers) which meet $r_j \theta + s_j \phi = l_j$ in an inner point of† $\mathscr{S}_j$. We select one such line, say $r_{j+1}\theta + s_{j+1}\phi = l_{j+1}$ with

$$\text{g.c.d.}\ (r_{j+1}, s_{j+1}, l_{j+1}) = 1 \quad \text{and} \quad t_{j+1} = \max(|r_{j+1}|, |s_{j+1}|) > t_j.$$

All points $(\theta_{j+1}, \phi_{j+1})$ on $r_{j+1}\theta + s_{j+1}\phi = l_{j+1}$ near enough to its intersection with $r_j \theta + s_j \phi = l_j$ thus satisfy simultaneously

$$|\theta_{j+1} - \theta_j| < \delta_j, \quad |\phi_{j+1} - \phi_j| < \delta_j, \tag{6}$$

$$|r_j \theta_{j+1} + s_j \phi_{j+1} - l_j| < \omega(t_{j+1}). \tag{7}$$

---

† For example if $\theta = a/c$, $\phi = b/c$ is a rational point on $r_j \theta + s_j \phi = l_j$ which is also an inner point of $\mathscr{S}_j$, where a, b, c are integers, then any solution of $ra + sb = lc$ will do.

We may suppose in addition that θ_{j+1} and ϕ_{j+1} are irrational. If r, s, l are any integers such that $0 < \max(|r|, |s|) < t_{j+1}$, the line $r\theta + s\phi = l$ cannot coincide with $r_{j+1}\theta + s_{j+1}\phi = l_{j+1}$ and so the two lines meet in a point with rational co-ordinates. Hence

$$\| r\theta_{j+1} + s\phi_{j+1} \| \neq 0, \quad 0 < \max(|r|, |s|) < t_{j+1}. \tag{8}$$

By continuity (6), (7), (8) continue to hold if θ, ϕ are substituted for θ_{j+1}, ϕ_{j+1} provided that

$$|\theta - \theta_{j+1}| \leqslant \delta_{j+1}, \quad |\phi - \phi_{j+1}| \leqslant \delta_{j+1},$$

and $\delta_{j+1} > 0$ is chosen sufficiently small. Hence for the constructed t_{j+1}, θ_{j+1}, ϕ_{j+1}, δ_{j+1} the statements (ii)$_{j+1}$, (iii)$_{j+1}$, (iv)$_{j+1}$ all hold. The statement (i)$_{j+1}$ holds for $t \leqslant t_j$ by (i)$_j$ and (iv)$_{j+1}$. If, however, $t_j < t \leqslant t_{j+1}$, by (7) with (θ, ϕ) instead of $(\theta_{j+1}, \phi_{j+1})$ we have

$$\| r_j\theta + s_j\phi \| < \omega(t_{j+1}) \leqslant \omega(t),$$
$$0 < \max(|r_j|, |s_j|) = t_j < t,$$

as required, since $\omega(t)$ was assumed monotone.

The form constructed in Theorem XIV enables us to show that Kronecker's theorem (Theorem IV of Chapter III) cannot be extended to any quantitative formulation, however weak, independent of the special forms in the way that the $\frac{1}{4}$ of Minkowski's theorem (Theorem II of Chapter III) is independent of θ provided that θ is irrational.

THEOREM XV. *Let $\epsilon(x) > 0$ $(x = 1, 2, \ldots)$ and let $\epsilon(x) \to 0$ as $x \to \infty$ (however slowly). Then there are numbers (θ, ϕ) such that $u\theta + v\phi \neq$ integer for integers $(u, v) \neq (0, 0)$ and numbers α, β such that*

$$\| \theta x - \alpha \| < \epsilon(|x|), \quad \| \phi x - \beta \| < \epsilon(|x|) \tag{9}$$

has only a finite number of integer solutions x.

Proof. As was shown in the proof of Theorem III there exist† numbers α, β such that

$$\| u\alpha + v\beta \| \geqslant \delta(\max(|u|, |v|))^{-2} \tag{10}$$

---

† The reader who wishes to avoid an appeal to the theory of algebraic numbers will have no difficulty in modifying the proof of Theorem XIV so that s_j is always odd. Put $\alpha = 0$, $\beta = \frac{1}{2}$. The proof of Theorem XV can be easily modified using $\| r_j\alpha + s_j\beta \| = \frac{1}{2}$ instead of (10). But our proof shows that α, β may be irrational.

for all integers $(u, v) \neq (0, 0)$, where $\delta > 0$. For integral t there is certainly some integer $X(t)$ such that

$$\epsilon(x) < \tfrac{1}{4}\delta t^{-3} \quad \text{for all} \quad x \geqslant X(t). \tag{11}$$

Write $$\omega(t) = \delta/2t^2 X(t+1), \tag{12}$$

and let θ, ϕ be the corresponding numbers constructed by Theorem XIV.

Since $\epsilon(x) \to 0$ there are at most a finite number of solutions of (9) with $\epsilon(|x|) \geqslant \tfrac{1}{4}\delta$. We shall show that there are none with $\epsilon(|x|) < \tfrac{1}{4}\delta$. Suppose that there is one and define the integer $t \geqslant 1$ by

$$t^3 \leqslant \delta/4\epsilon(|x|) < (t+1)^3. \tag{13}$$

Then, by (11) and (12)

$$|x| < X(t+1) = \delta/2t^2\omega(t). \tag{14}$$

By hypothesis we may find integers (u, v) such that

$$\| u\theta + v\phi \| < \omega(t), \quad 0 < \max(|u|, |v|) \leqslant t.$$

By (9) and (10) we have

$$\begin{aligned}
\delta t^{-2} &\leqslant \| u\alpha + v\beta \| \\
&= \| u(\alpha - \theta x) + v(\beta - \phi x) + x(u\theta + v\phi) \| \\
&\leqslant |u| \, \| \alpha - \theta x \| + |v| \, \| \beta - \phi x \| + |x| \, \| u\theta + v\phi \| \\
&< 2t\epsilon(|x|) + |x|\, \omega(t). \tag{15}
\end{aligned}$$

By (13) and (14) the two summands at the end of (15) are both $\leqslant \tfrac{1}{2}\delta t^{-2}$. This contradiction proves the theorem.

8. A quantitative Kronecker's theorem.

We first prove the more general

THEOREM XVI. *Let* $f_k(\mathbf{z})$, $g_k(\mathbf{w})$ *for* $1 \leqslant k \leqslant l$ *be linear forms in* $\mathbf{z} = (z_1, \ldots, z_l)$, $\mathbf{w} = (w_1, \ldots, w_l)$ *respectively. Suppose that*

$$\sum_k f_k(\mathbf{z}) g_k(\mathbf{w}) = \sum_k z_k w_k \tag{1}$$

identically. Let $\boldsymbol{\beta} = (\beta_1, \ldots, \beta_l)$ *be any real numbers.*

A. *A necessary condition that*

$$|\beta_k - f_k(\mathbf{b})| \leqslant 1 \quad (1 \leqslant k \leqslant l), \tag{2}$$

CDA

for some integral **b** *is that*

$$\| \Sigma g_k(\mathbf{w}) \beta_k \| \leqslant l \max | g_k(\mathbf{w}) | \qquad (3)$$

for all integral **w**.

B. *A sufficient condition that* (2) *holds for some integral* **b** *is that*

$$\| \Sigma g_k(\mathbf{w}) \beta_k \| \leqslant 2^{l-1}(l!)^{-2} \max | g_k(\mathbf{w}) | \qquad (4)$$

for all integral **w**.

Proof of A. $\Sigma f_k(\mathbf{b}) g_k(\mathbf{w}) = \Sigma w_k b_k$ is an integer if **w**, **b** are, so (2) implies

$$\| \Sigma g_k(\mathbf{w}) \beta_k \| = \| \Sigma g_k(\mathbf{w}) (\beta_k - f_k(\mathbf{b})) \|$$
$$\leqslant \Sigma | g_k(\mathbf{w}) | \leqslant l \max | g_k(\mathbf{w}) |.$$

Proof of B. We regard **w** as a row matrix and **z**, **β** as column matrices. Let **G** be the square matrix whose kth column is the coefficients of g_k and **F** the square matrix whose kth row is the coefficients of f_k. Then (1) becomes

$$\mathbf{G} = \mathbf{F}^{-1}. \qquad (5)$$

By Theorem VI and Lemma 4 of Appendix B applied to the region $\max | g_j(\mathbf{w}) | \leqslant 1$, there is an integral $l \times l$ matrix **W** with $\det \mathbf{W} = 1$ whose kth row $\mathbf{w}^{(k)}$ satisfies

$$\max_{j} | g_j(\mathbf{w}^{(k)}) | = \mu_k, \quad \prod_k \mu_k \leqslant 2^{1-l} . l! | \det \mathbf{G} |. \qquad (6)$$

Now **WGβ** is a column matrix with kth element $\sum_j \beta_j g_j(\mathbf{w}^{(k)})$. Hence by (4), (6)

$$\mathbf{WG\beta} = \mathbf{a} + \boldsymbol{\delta},$$

where **a** is integral and

$$\max | \delta_k | \leqslant 2^{l-1}(l!)^{-2} \mu_k. \qquad (7)$$

Hence, by (5), $$\boldsymbol{\beta} = \mathbf{Fb} + \boldsymbol{\gamma}, \qquad (8)$$

where $$\mathbf{b} = \mathbf{W}^{-1}\mathbf{a}; \quad \boldsymbol{\delta} = \mathbf{WG\gamma}. \qquad (9)$$

Here **b** is integral since $\det \mathbf{W} = 1$. By Cramer's rule γ_j is $(\det \mathbf{G})^{-1}$ times the determinant of the matrix obtained from **WG** by substituting **δ** for the jth column. But the elements of the kth row of **WG** are at the most μ_k by (6); and so, on estimating the determinant crudely using (7),

$$| \gamma_j | \leqslant | \det \mathbf{G} |^{-1} . l! . 2^{l-1}(l!)^{-2} \prod \mu_k$$
$$\leqslant 1 \text{ by (6).} \qquad (10)$$

But (8), (10) are just (2).

THEOREM XVII. *Let $L_j(\mathbf{x})$, $M_i(\mathbf{u})$ be as defined in § 1 and $l = m+n$. Let $\boldsymbol{\alpha} = (\alpha_1, \ldots, \alpha_n)$, $C > 0$, $X > 1$ be given.*

A. *A necessary condition that*

$$\| L_j(\mathbf{a}) - \alpha_j \| \leqslant C, \quad |a_i| \leqslant X, \tag{11}$$

for some integral $\mathbf{a}$ is that

$$\| \mathbf{u\alpha} \| \leqslant \gamma \max (X \max \| M_i(\mathbf{u}) \|, C \max |u_j|), \tag{12}$$

hold for all integral $\mathbf{u}$ with $\gamma = l$.

B. *A sufficient condition that* (11) *be soluble is that* (12) *hold for all integral $\mathbf{u}$ with $\gamma = 2^{l-1} (l!)^{-2}$.*

Proof. Special case of Theorem XVI with

$$\mathbf{z} = (\mathbf{x}, \mathbf{y}) = (x_1, \ldots, x_m, y_1, \ldots, y_n);$$

$$\mathbf{w} = (\mathbf{v}, \mathbf{u}) = (v_1, \ldots, v_m, u_1, \ldots, u_n);$$

$$f_k(\mathbf{z}) = \begin{cases} C^{-1}(L_k(\mathbf{x}) + y_k) & \text{for} \quad k \leqslant n, \\ X^{-1} x_{k-n} & \text{for} \quad n < k \leqslant l; \end{cases}$$

$$g_k(\mathbf{w}) = \begin{cases} Cu_k & \text{for} \quad k \leqslant n, \\ X(v_{k-n} - M_{k-n}(\mathbf{u})) & \text{for} \quad n < k \leqslant l; \end{cases}$$

and $\qquad \boldsymbol{\beta} = (C^{-1}\boldsymbol{\alpha}, 0).$

We now deduce Kronecker's theorem (Theorem IV of Chapter III) from Theorem XVII B. In the present notation it states that if $\| \mathbf{u\alpha} \| = 0$ with integral $\mathbf{u}$ whenever

$$\| M_i(\mathbf{u}) \| = 0 \quad (1 \leqslant i \leqslant m),$$

then for any $\epsilon > 0$ there is an integral $\mathbf{a}$ with

$$\| L_j(\mathbf{a}) - \alpha_j \| < \epsilon \quad (1 \leqslant j \leqslant n).$$

Put $C = \epsilon$. Since $\| \mathbf{u\alpha} \| \leqslant \frac{1}{2}$, the condition (12) is satisfied except for the $\mathbf{u}$ with $\max |u_j| \leqslant \frac{1}{2}\gamma^{-1}\epsilon^{-1}$. But, by hypothesis, we may choose X so large that (12) holds for the finitely many remaining $\mathbf{u}$. Thus Theorem XVII B applies.

9. Successive minima.†

The use of the successive minima of a convex body makes the formal relations between the transference theorems of §§ 1-4 clearer but leads to less precise results, as we shall indicate briefly.

† This may be omitted at first reading.

Let $f_k(\mathbf{z})$ be l linear forms in $\mathbf{z} = (z_1, \ldots, z_l)$ of determinant $\Delta \neq 0$. Then $F(\mathbf{z}) = \max |f_k(\mathbf{z})|$ is the distance function of the convex body $|f_k(\mathbf{z})| \leqslant 1$ $(1 \leqslant k \leqslant l)$ of volume $2^l |\Delta|^{-1}$. The successive minima $\lambda_1, \ldots, \lambda_l$ satisfy

$$\lambda_1 \leqslant \lambda_2 \leqslant \ldots \leqslant \lambda_l, \quad (l!)^{-1} |\Delta| \leqslant \Pi \lambda_j \leqslant |\Delta|, \tag{1}$$

by Theorem V of Appendix B.

Put
$$\Lambda = \sup_{\zeta} (\inf_{\mathbf{z}} F(\zeta - \mathbf{z})),$$

where ζ, $\mathbf{z}$ run through all vectors and all integer vectors respectively. Since the lower bound is attained as we saw earlier (p. 81), Λ is the least number such that for any ζ there is an integer $\mathbf{z}$ with $F(\zeta - \mathbf{z}) \leqslant \Lambda$. Our first aim is

$$\lambda_l \leqslant 2\Lambda \leqslant \lambda_1 + \ldots + \lambda_l \quad (\leqslant l\lambda_l). \tag{2}$$

There is an integral $\mathbf{x}^{(k)}$ such that $F(\zeta^{(k)} - \mathbf{x}^{(k)}) \leqslant \Lambda$, where $\zeta^{(k)}$ has $\frac{1}{2}$ in the kth place but 0 elsewhere. Thus $F(\mathbf{y}^{(k)}) \leqslant 2\Lambda$ where $\mathbf{y}^{(k)} = 2(\zeta^{(k)} - \mathbf{x}^{(k)})$ has an odd kth element but all other elements even. The $\mathbf{y}^{(k)}$ must be linearly independent since their determinant is clearly odd. The left part of (2) now follows from the definition of λ_l. Denote by $\mathbf{z}^{(k)}$ linearly independent integer vectors with $F(\mathbf{z}^{(k)}) = \lambda_k$. Any ζ may be written

$$\zeta = \beta_1 \mathbf{z}^{(1)} + \ldots + \beta_l \mathbf{z}^{(l)}.$$
Let
$$\mathbf{z} = b_1 \mathbf{z}^{(1)} + \ldots + b_l \mathbf{z}^{(l)},$$

where the b_k are integers and $|b_k - \beta_k| \leqslant \frac{1}{2}$. Then

$$\begin{aligned}
F(\zeta - \mathbf{z}) &= F(\Sigma(\beta_k - b_k) \mathbf{z}^{(k)}) \\
&\leqslant \Sigma |\beta_k - b_k| F(\mathbf{z}^{(k)}) \\
&\leqslant \tfrac{1}{2}\Sigma \lambda_k.
\end{aligned}$$

This gives the right-hand side of (2).

It is readily verified that the maximum of the right-hand side of (2) subject to $\lambda_1 \geqslant 1$ and (1) is $l - 1 + |\Delta|$. This is rather weaker than Theorem V (which then gives $2\Lambda \leqslant [\Delta] + 1$). On the other hand

$$\lambda_1 \lambda_l^{l-1} \geqslant (l!)^{-1} |\Delta| \tag{3}$$

by (1). This with (2) gives an estimate for Λ from below when $\lambda_1 \geqslant 1$ similar to, but weaker than, that of § 5.

Let now $g_k(\mathbf{w})$ be forms such that identically

$$\Sigma f_k(\mathbf{z})\, g_k(\mathbf{w}) = \Sigma z_k w_k. \tag{4}$$

Let $\mu_1, \ldots, \mu_l$ be the successive minima of $G(\mathbf{w}) = \max |g_k(\mathbf{w})|$. We shall show that

$$l^{-1} \leqslant \lambda_k \mu_{l+1-k} \leqslant (l-1)!. \tag{5}$$

We adopt the conventions of the proof of Theorem XVI B; in particular (8·5) holds. Let $\mathbf{Z}$ be an integer matrix whose columns are the $\mathbf{z}^{(k)}$. Identically

$$(\text{adj}\,\mathbf{Z})\,\mathbf{G} = \Delta^{-1} \text{adj}\,(\mathbf{FZ}), \tag{6}$$

where 'adj' denotes the adjoint matrix, i.e. the transposed matrix of cofactors. The elements of the kth column of $\mathbf{FZ}$ are of the type $f_i(\mathbf{z}^{(k)})$, and so at most λ_k in absolute value. The elements of the kth row of adj $(\mathbf{FZ})$ are thus at most†

$$(l-1)! \prod_{j \neq k} \lambda_j \leqslant (l-1)! \,|\Delta|\, \lambda_k^{-1}, \tag{7}$$

on estimating crudely and using (1). Hence by (6),

$$|g_j(\overline{\mathbf{w}}^{(k)})| \leqslant (l-1)!\, \lambda_k^{-1},$$

where $\overline{\mathbf{w}}^{(k)}$ is the kth row of adj $\mathbf{Z}$. There are thus $l+1-k$ linearly independent integer $\mathbf{w}$ with $G(\mathbf{w}) \leqslant (l-1)!\, \lambda_k^{-1}$, which proves the right-hand side of (5). Let‡ now $\mathbf{w}^{(k)}$ be linearly independent integer vectors with $G(\mathbf{w}^{(k)}) = \mu_k$. The $\mathbf{z}$ with

$$\mathbf{w}^{(j)}\mathbf{z} = 0 \quad (1 \leqslant j \leqslant l+1-k),$$

lie in a subspace of dimension $k-1$, and so there exist i, j with

$$\mathbf{w}^{(j)}\mathbf{z}^{(i)} \neq 0, \quad 1 \leqslant i \leqslant k; \ 1 \leqslant j \leqslant l+1-k.$$

Hence, since $\mathbf{w}^{(j)}, \mathbf{z}^{(i)}$ are integral,

$$1 \leqslant |\mathbf{w}^{(j)}\mathbf{z}^{(i)}| = \left| \sum_h f_h(\mathbf{z}^{(i)})\, g_h(\mathbf{w}^{(j)}) \right|$$

$$\leqslant l F(\mathbf{z}^{(i)})\, G(\mathbf{w}^{(j)})$$

$$= l\lambda_i \mu_j \leqslant l\lambda_k \mu_{l+1-k},$$

which is the left of (5).

---

† Lemma 5, Corollary, of Chapter VIII permits the replacement of $(l-1)!$ by $(l-1)^{\frac{1}{2}(l-1)}$ in (7).

‡ This departs from the notation of § 8.

Condition (4) is rather stronger than the condition in Theorem I that $\Sigma f_k(\mathbf{z}) g_k(\mathbf{w})$ have integral coefficients but it covers all the applications made. By (3), (5) we have $\mu_1 \leqslant \gamma_1 \mid \lambda_1 d \mid^{1/(d-1)}$ where $d = \Delta^{-1}$ and γ_1 depends only on l. Apart from the value of γ_1 this is Theorem I for our special case (4). Further, (2), (4) together give $0 < \gamma_3 \leqslant \mu_1 \Lambda \leqslant \gamma_2$ with γ_2, γ_3 depending only on l, which relates the homogeneous problem for the g_k to the inhomogeneous problem for the f_k.

NOTES

§ 2. MAHLER (1939a). For a generalization to convex bodies see MAHLER (1939b).

§ 3. For a more precise form of Theorem III see DAVENPORT (1954), (1955). For a proof that there is a $\gamma > 0$ such that (3·1) is true for non-enumerably many real sets θ_j and all integers $x > 0$ see CASSELS (1955).

§ 4. HLAWKA (1952). Clearly Theorem V generalizes almost at once to all convex bodies. For more precise results see KNESER (1955) and BIRCH (1956).

§ 5. BIRCH (1957); where more precise results are given.

§ 6. KHINTCHINE (1948b) and CASSELS (1952b). For generalizations, see CASSELS (1952b), CHABAUTY & LUTZ (1950) and HLAWKA (1954b), and for related work see JARNÍK (1946), (1954). The value of the best constant in Theorem XI is an interesting unsolved problem. It can be shown to be $\leqslant 1/12$ and $> 1/45·2$.

For a form of Lemma 2 valid for all $k > 1$ see KHINTCHINE (1926).

§ 7. KHINTCHINE (1926) and (1948b).

§ 8. KHINTCHINE (1948a). The argument here was suggested by Mr B. J. Birch.

§ 9. Most of the arguments here go back to Mahler. (E.g. MAHLER (1955).)

CHAPTER VI

RATIONAL APPROXIMATION TO ALGEBRAIC NUMBERS. ROTH'S THEOREM

1. Introduction. This chapter requires absolutely no prior knowledge of the theory of algebraic numbers. An ALGEBRAIC NUMBER ξ is one which satisfies an equation

$$f(\xi) = 0, \quad f(x) = a_n x^n + a_{n-1} x^{n-1} + \ldots + a_0, \tag{1}$$

where $a_n, \ldots, a_0$ are rational. On multiplying $f(x)$ by a suitable integer we may suppose that $a_n, \ldots, a_0$ are integers and, without loss of generality,† that $a_n \neq 0$. As Liouville first remarked, an irrational algebraic number ξ cannot be too closely approximated by rationals. His argument is simple.‡ Let $\xi = \xi_1, \xi_2, \ldots, \xi_n$ be the roots of $f(x) = 0$, so that $f(x) = a_n \Pi(x - \xi_j)$. Suppose that $q > 0$, p are integers and that $|\xi - p/q| < 1$. Then on the one hand

$$\begin{aligned}
f(p/q) &= |a_n| \, |\xi - p/q| \prod_{j \geqslant 2} |\xi_j - p/q| \\
&\leqslant |a_n| \, |\xi - p/q| \prod_{j \geqslant 2} (|\xi| + 1 + |\xi_j|) \\
&= c \, |\xi - p/q|,
\end{aligned} \tag{2}$$

where $c > 0$ is a constant. On the other hand

$$q^n f(p/q) = a_n p^n + a_{n-1} p^{n-1} q + \ldots + a_0 q^n$$

is an integer, and so

$$|q^n f(p/q)| \geqslant 1, \tag{3}$$

apart from the finite number of possibilities $p/q = \xi_j \neq \xi$ ($p/q = \xi$ is impossible since ξ is irrational). Comparison of (2) and (3) shows that there are only a finite number of solutions of

$$|\xi - p/q| < c^{-1} q^{-n}. \tag{4}$$

This chapter is devoted to proving the much stronger

---

† We do not assume that $f(x)$ is irreducible. As the concept of irreducibility is not needed later in the chapter we do not introduce it here.

‡ Cf. Theorem III of Chapter V.

THEOREM I. (Roth.) *Let ξ be an irrational algebraic number and $\delta > 0$ be arbitrarily small. Then there are only a finite number of pairs of integers $q > 0$, p such that*

$$|\xi - p/q| < q^{-2-\delta}. \tag{5}$$

Note that the degree n does not occur in the enunciation. Since, as we saw in Chapter I, there are infinitely many solutions $q > 0$, p of (5) with $\delta = 0$ for any irrational ξ, this theorem is the best possible of its kind. But for $n = 2$ Liouville's result about (4) is yet stronger.

2. Preliminaries. We first show that we need prove the theorem only when† $a_n = 1$ in (1·1). For $a_n \xi = \Xi$ satisfies $\Xi^n + a_{n-1} \Xi^{n-1} + \ldots + a_n^{n-1} a_0 = 0$. If (1·5) holds then

$$|\Xi - a_n p/q| < |a_n| \, q^{-2-\delta} < q^{-2-\frac{1}{2}\delta}, \tag{1}$$

if q is large enough; and so (1) has infinitely many solutions if (1·5) has. Since δ is arbitrary, Ξ would contradict the theorem if ξ does. We shall therefore assume that

$$f(\xi) = 0, \quad f(x) = x^n + a_{n-1} x^{n-1} + \ldots + a_0, \tag{2}$$

for integers $a_{n-1}, \ldots, a_0$. We put

$$a = \max(1, |a_{n-1}|, \ldots, |a_0|). \tag{3}$$

This meaning of $\xi, f(x), n, a$ will be standard for the rest of the chapter.

We shall use polynomials of the type

$$R(x_1, \ldots, x_m) = \sum_{\substack{0 \leqslant j_\mu \leqslant r_\mu \\ (1 \leqslant \mu \leqslant m)}} C(j_1, \ldots, j_m) \, x_1^{j_1} \ldots x_m^{j_m},$$

in the large number m of variables x_μ ($1 \leqslant \mu \leqslant m$) with real coefficients $C(j_1, \ldots, j_m)$. We write

$$\lceil R \rceil = \max |C(j_1, \ldots, j_m)|,$$

and
$$R_{i_1 \ldots i_m} = \frac{1}{i_1! \ldots i_m!} \frac{\partial^{i_1 + \ldots + i_m} R}{\partial x_1^{i_1} \ldots \partial x_m^{i_m}},$$

for any non-negative integers i_μ.

---

† I.e. when ξ is an algebraic integer.

LEMMA 1. *If R has integer coefficients so has $R_{i_1\ldots i_m}$. If R has degree r_μ in x_μ then $R_{i_1\ldots i_m}$ has degree at most $r_\mu - i_\mu$ (and so vanishes if $i_\mu > r_\mu$ for any μ). Finally,*

$$\lceil R_{i_1\ldots i_m}\rceil \leqslant 2^{r_1+\cdots+r_m}\lceil R\rceil.$$

Proof. We have

$$R_{i_1\ldots i_m} = \sum_{i_\mu \leqslant j_\mu \leqslant r_\mu}\binom{j_1}{i_1}\cdots\binom{j_m}{i_m}C(j_1,\ldots,j_m)x_1^{j_1-i_1}\ldots x_m^{j_m-i_m},$$

$$(4)$$

where the binomial coefficients $\binom{j}{i}$ are integers. Since

$$\binom{j}{i} \leqslant \sum_{0\leqslant i\leqslant j}\binom{j}{i} = (1+1)^j \leqslant 2^r \qquad (5)$$

if $0\leqslant i\leqslant j\leqslant r$, the result follows.

By Taylor's theorem we have the identity

$$R(x_1+y_1,\ldots,x_m+y_m) = \sum_{0\leqslant i_\mu\leqslant r_\mu} y_1^{i_1}\ldots y_m^{i_m}R_{i_1\ldots i_m}(x_1,\ldots,x_m).$$

$$(6)$$

We shall say that R has THE INDEX I AT $(\alpha_1,\ldots,\alpha_m)$ WITH RESPECT TO $(s_1,\ldots,s_m)$, where $\alpha_1,\ldots,\alpha_m$ are any numbers and $s_1,\ldots,s_m$ are positive integers, if I is the least value of $\Sigma i_\mu/s_\mu$ for which $R_{i_1\ldots i_m}(\alpha_1,\ldots,\alpha_m)$ does not vanish. From (6) it follows that such $i_1,\ldots,i_m$ exist except when R vanishes identically; and then we say, conventionally, that the index is $+\infty$.

LEMMA 2. *If* ind *denotes the index at $(\alpha_1,\ldots,\alpha_m)$ with respect to $(s_1,\ldots,s_m)$ then*

(i) $\operatorname{ind} R_{i_1\ldots i_m} \geqslant \operatorname{ind} R - \Sigma i_\mu/s_\mu$;
(ii) $\operatorname{ind}(R^{(1)}+R^{(2)}) \geqslant \min(\operatorname{ind} R^{(1)}, \operatorname{ind} R^{(2)})$;
(iii) $\operatorname{ind} R^{(1)}R^{(2)} = \operatorname{ind} R^{(1)} + \operatorname{ind} R^{(2)}$.

Proof (i). Clear.

(ii), (iii) Put $s=s_1\ldots s_m$ and $I=\operatorname{ind} R$. By (6) clearly t^{sI} is the least power of t actually occurring in

$$R(\alpha_1+t^{s/s_1}y_1,\ldots,\alpha_m+t^{s/s_m}y_m),$$

considered as a polynomial in the independent variables

$$t, y_1,\ldots,y_m.$$

The proof of Theorem I falls into three main parts which will be considered separately in §§ 3, 4, 5. The conclusions of each section will be enunciated as Theorems II, III, IV respectively. Finally, in § 6 the proof of Theorem I will be rapid, using Theorems II, III, IV but nothing else out of §§ 3, 4, 5.

3. Construction of $R(x_1, \ldots, x_m)$.

THEOREM II. *Let $\epsilon > 0$ be arbitrary, let*

$$m > 8n^2\epsilon^{-2} \tag{1}$$

be an integer, n being the degree of $f(x)$, and let $r_1, \ldots, r_m$ be any positive integers. There is a polynomial $R(x_1, \ldots, x_m)$ with integer coefficients of degree at most r_μ in x_μ $(1 \leqslant \mu \leqslant m)$ which (i) does not vanish identically, (ii) has index at least

$$\tfrac{1}{2}m(1 - \epsilon) \tag{2}$$

at $(\xi, \ldots, \xi)$ with respect to $(r_1, \ldots, r_m)$, and (iii) satisfies

$$\boxed{R} \leqslant \gamma^{r_1 + \cdots + r_m}, \quad \gamma = 4(a+1), \tag{3}$$

where a is defined by (2·3).

Note. The shape of (1) and the value of γ are irrelevant. It is enough for our later purposes that the conclusions of the theorem hold for all m larger than some constant depending on ξ, ϵ and with some γ which may depend on ϵ, ξ, m.

The proof requires several lemmas.

LEMMA 3. *Let*

$$L_j = \sum_{1 \leqslant k \leqslant N} a_{jk} z_k \quad (1 \leqslant j \leqslant M),$$

be M linear forms with integer coefficients in $N > M$ variables. Suppose that

$$|a_{jk}| \leqslant A \quad (1 \leqslant j \leqslant M; \; 1 \leqslant k \leqslant N).$$

Then there exist integer values of the variables $z_1, \ldots, z_N$ not all 0 such that

$$L_j = 0 \quad (1 \leqslant j \leqslant M), \quad |z_i| \leqslant Z = [(NA)^{M/(N-M)}] \quad (1 \leqslant k \leqslant n).$$

Proof. We have

$$NA < (Z+1)^{(N-M)/M},$$

and hence $\quad NAZ + 1 \leqslant NA(Z+1) < (Z+1)^{N/M}.$

For any set of integer values of $\mathbf{z} = (z_1, \ldots, z_N)$ in

$$0 \leqslant z_k \leqslant Z \quad (1 \leqslant k \leqslant N), \tag{4}$$

we have $\quad -B_j Z \leqslant L_j(\mathbf{z}) \leqslant C_j Z, \quad B_j + C_j \leqslant NA,$

where $-B_j$, C_j are the sums of the negative and positive coefficients in $L_j(\mathbf{z})$ respectively. Hence the integer $L_j(\mathbf{z})$ may take at most $NAZ + 1$ values. There are thus $(Z+1)^N$ values of $\mathbf{z}$ but only $(NAZ+1)^M < (Z+1)^N$ sets of values of $(L_1, \ldots, L_M)$ that may arise from them. There must therefore be two distinct vectors $\mathbf{z}^{(1)}$, $\mathbf{z}^{(2)}$ in (4) such that $L_j(\mathbf{z}^{(1)}) = L_j(\mathbf{z}^{(2)})$ $(1 \leqslant j \leqslant M)$. Now $\mathbf{z} = \mathbf{z}^{(1)} - \mathbf{z}^{(2)}$ clearly satisfies the conclusions of the lemma.

LEMMA 4. *For each integer $l \geqslant 0$ there are rational integers $a_j^{(l)}$ $(0 \leqslant j < n)$ such that*

$$\xi^l = a_{n-1}^{(l)} \xi^{n-1} + \ldots + a_0^{(l)},$$

and, for the a of (2·3),

$$|a_j^{(l)}| \leqslant (a+1)^l.$$

Proof. For $l < n$ trivial. For $l \geqslant n$ follows by induction since

$$\xi^l = \xi \cdot \xi^{l-1} = a_{n-1}^{(l-1)} \xi^n + \ldots + a_0^{(l-1)} \xi,$$

and $\xi^n = -a_{n-1} \xi^{n-1} - \ldots - a_0$.

LEMMA 5. *For any positive integers $r_1, \ldots, r_m$ and real $\lambda > 0$ the number of sets of integers $i_1, \ldots, i_m$ such that*

$$\Sigma i_\mu / r_\mu \leqslant \tfrac{1}{2}(m - \lambda), \quad 0 \leqslant i_\mu \leqslant r_\mu \quad (1 \leqslant \mu \leqslant m)$$

is at most $\quad (2m)^{\frac{1}{2}} \lambda^{-1} (r_1 + 1) \ldots (r_m + 1). \tag{5}$

Proof. Trivial for $m = 1$ since the number of solutions is at most $r_1 + 1$, and is 0 if $\lambda > 1$.

For $m > 1$ we may assume that

$$\lambda > (2m)^{\frac{1}{2}} > 1, \tag{6}$$

since otherwise the lemma is trivial, and also that the lemma has already been proved for $m - 1$. Hence for fixed $r = r_m$, $i = i_m$ the number of integers $i_1, \ldots, i_{m-1}$ is at most

$$(2m-2)^{\frac{1}{2}} (\lambda - 1 + 2i/r)^{-1} (r_1 + 1) \ldots (r_{m-1} + 1). \tag{7}$$

But
$$\sum_{0 \leqslant i \leqslant r} \frac{2}{\lambda - 1 + 2i/r} = \Sigma\left(\frac{1}{\lambda - 1 + 2i/r} + \frac{1}{\lambda + 1 - 2i/r}\right)$$

$$= \Sigma \frac{2\lambda}{\lambda^2 - (1 - 2i/r)^2}$$

$$< 2(r+1)\lambda/(\lambda^2 - 1), \tag{8}$$

and, by (6),
$$\lambda^2 - 1 > \lambda^2(1 - 1/2m) > \lambda^2(1 - 1/m)^{\frac{1}{2}}. \tag{9}$$

The truth of the lemma now follows from (7), (8), (9) on allowing $i = i_m$ to take all values $0, 1, \ldots, r = r_m$.

Proof of Theorem II. We write
$$R(x_1, \ldots, x_m) = \sum_{0 \leqslant j_\mu \leqslant r_\mu} C(j_1, \ldots, j_m)\, x_1^{j_1} \ldots x_m^{j_m},$$

where the $C(j_1, \ldots, j_m)$ are
$$N = (r_1 + 1) \ldots (r_m + 1)$$

integers to be determined. We must have
$$R_{i_1 \ldots i_m}(\xi, \ldots, \xi) = 0, \tag{10}$$

for all integers $i_1, \ldots, i_m$ such that
$$\Sigma i_\mu / r_\mu \leqslant \tfrac{1}{2}m(1 - \epsilon). \tag{11}$$

Since (10) is trivially true if $i_\mu > r_\mu$ for any μ, we may suppose that
$$0 \leqslant i_\mu \leqslant r_\mu \quad (1 \leqslant \mu \leqslant m). \tag{12}$$

On expressing all the powers of ξ in terms of $1, \xi, \ldots, \xi^{n-1}$ by Lemma 4, we see that (10) is implied by† n linear conditions on the $C(j_1, \ldots, j_m)$ with rational integer coefficients. By (2·4) and Lemma 4, the coefficients are of the type
$$\binom{j_1}{i_1} \ldots \binom{j_m}{i_m} a_j^{(l)} \quad (0 \leqslant j < n), \tag{13}$$

where
$$l = (j_1 - i_1) + \ldots + (j_m - i_m) \leqslant r_1 + \ldots + r_m.$$

Hence (13) are integers at most
$$A = (2a + 2)^{r_1 + \ldots + r_m} \tag{14}$$

---

† (10) is equivalent to the n conditions if $f(x)$ is irreducible, but we need not assume this.

in absolute value by Lemma 4 and by (2·5). By Lemma 5 with $\lambda = m\epsilon$ and by (1) the number M of conditions is

$$M \leqslant n \, . \, (2m)^{\frac{1}{2}} (m\epsilon)^{-1} N \leqslant \tfrac{1}{2}N. \tag{15}$$

By Lemma 3 there exist integers $C(j_1, \ldots, j_m)$ not all 0 satisfying (10) for (11), (12) and such that

$$|C(j_1, \ldots, j_m)| \leqslant (NA)^{M/(N-M)}$$

$$\leqslant NA$$

$$\leqslant \gamma^{r_1 + \cdots + r_m},$$

by (14), (15) and since, by (2·5),

$$N = (r_1 + 1) \ldots (r_m + 1) \leqslant 2^{r_1 + \cdots + r_m}.$$

4. Behaviour of R at rational points near $(\xi, \ldots, \xi)$.

THEOREM III. *Let $q_\mu > 0, p_\mu$ $(1 \leqslant \mu \leqslant m)$ be integers and let*

$$\eta_\mu = \frac{p_\mu}{q_\mu} - \xi, \quad |\eta_\mu| < q_\mu^{-2-\delta}, \tag{1}$$

where $$0 < \delta < 1/12. \tag{2}$$

Let ϵ be any number such that

$$0 < \epsilon < \delta/20, \tag{3}$$

$$q_\mu^\epsilon > 64 \, (a+1) \max (1, |\xi|) \quad (1 \leqslant \mu \leqslant m). \tag{4}$$

Let $r_1, \ldots, r_m$ be any positive integers such that

$$r_1 \log q_1 \leqslant r_\mu \log q_\mu \leqslant (1+\epsilon) r_1 \log q_1 \quad (1 \leqslant \mu \leqslant m). \tag{5}$$

Then the index of the R constructed in Theorem II at

$$(p_1/q_1, \ldots, p_m/q_m),$$

with respect to $(r_1, \ldots, r_m)$ is at least

$$\delta m/8. \tag{6}$$

Note. The shape of (2)–(6) is immaterial. It is enough that some explicit lower bound for the index should exist valid if the

q_μ are large enough and the $r_\mu \log q_\mu$ substantially equal to each other.

Proof. Let $j_1, ..., j_m$ be any non-negative integers such that

$$\Sigma j_\mu / r_\mu < \delta m/8, \tag{7}$$

and put $\quad T(x_1, ..., x_m) = R_{j_1 ... j_m}(x_1, ..., x_m).$

We must show that

$$T(p_1/q_1, ..., p_m/q_m) = 0.$$

By Theorem II and Lemma 1, $\overline{|T|} \leqslant (2\gamma)^{r_1 + \cdots + r_m}$ and T has integer coefficients. Since T has degree at most r_μ in x_μ it contains at most $(r_1 + 1) \ldots (r_m + 1) \leqslant 2^{r_1 + \cdots + r_m}$ terms. Hence for any positive integers $i_1, ..., i_m$ we have, using Lemma 1 again and estimating crudely, that

$$|T_{i_1 ... i_m}(\xi, ..., \xi)|$$
$$\leqslant (r_1 + 1) \ldots (r_m + 1) . 2^{r_1 + \cdots + r_m} . (2\gamma)^{r_1 + \cdots + r_m} (\max(1, |\xi|))^{r_1 + \cdots + r_m}$$
$$\leqslant \gamma_1^{r_1 + \cdots + r_m} \text{ (say)}, \quad \gamma_1 = 8\gamma \max(1, |\xi|). \tag{8}$$

By Lemma 2, Theorem II (ii), (3) and (7) the index of T at $(\xi, ..., \xi)$ with respect to $(r_1, ..., r_m)$ is at least

$$\tfrac{1}{2}m(1 - \epsilon) - \Sigma j_\mu / r_\mu > \tfrac{1}{2}m(1 - \epsilon - \tfrac{1}{4}\delta) > \tfrac{1}{2}m(1 - \tfrac{1}{3}\delta). \tag{9}$$

But by (2·6) and (1), we have

$$T(p_1/q_1, ..., p_m/q_m) = \sum_{0 \leqslant i_\mu \leqslant r_\mu} T_{i_1 ... i_m}(\xi, ..., \xi) \eta_1^{i_1} ... \eta_m^{i_m}, \tag{10}$$

where, by (9), the summand vanishes unless

$$\Sigma i_\mu / r_\mu \geqslant \tfrac{1}{2}m(1 - \tfrac{1}{3}\delta). \tag{11}$$

For such $i_1, ..., i_m$ the inequalities (1), (5) give

$$-\log|\eta_1^{i_1} ... \eta_m^{i_m}| \geqslant (2 + \delta) \Sigma i_\mu \log q_\mu$$
$$\geqslant (2 + \delta) r_1 \log q_1 \Sigma i_\mu / r_\mu$$
$$\geqslant (2 + \delta) r_1 \log q_1 . \tfrac{1}{2}m(1 - \tfrac{1}{3}\delta)$$
$$\geqslant (1 + \tfrac{1}{2}\delta)(1 - \tfrac{1}{3}\delta)(1 + \epsilon)^{-1} \Sigma r_\mu \log q_\mu.$$

But $\quad (1 + \tfrac{1}{2}\delta)(1 - \tfrac{1}{3}\delta) = 1 + \tfrac{1}{6}\delta(1 - \delta) > 1 + \tfrac{1}{8}\delta > (1 + \epsilon)^2$

by (2), (3) and so, finally,

$$|\eta_1^{i_1} ... \eta_m^{i_m}| < (q_1^{r_1} ... q_m^{r_m})^{-1-\epsilon}. \tag{12}$$

Since there are at most $(r_1+1)\dots(r_m+1)\leqslant 2^{r_1+\dots+r_m}$ terms in (10) we deduce from (8), (10), (12) that

$$|q_1^{r_1}\dots q_m^{r_m}T(p_1/q_1,\dots,p_m/q_m)|$$
$$<\prod_\mu(2\gamma_1 q_\mu^{-\epsilon})^{r_\mu}<1$$

by (4), since $2\gamma_1=16\gamma\max(1,|\xi|)=64(a+1)\max(1,|\xi|)$ by (3·3) and (8). But $q_1^{r_1}\dots q_m^{r_m}T(p_1/q_1,\dots,p_m/q_m)$ is an integer. It must be 0; which proves the theorem.

5. Behaviour at rational points of a polynomial with integer coefficients.

THEOREM IV. *Let*

$$\omega=\omega(m,\epsilon)=24\cdot 2^{-m}(\epsilon/12)^{2^{m-1}}, \tag{1}$$

where m is a positive integer and

$$0<\epsilon<1/12. \tag{2}$$

Let $r_1,\dots,r_m$ be positive integers such that

$$\omega r_\mu\geqslant r_{\mu+1}\quad(1\leqslant\mu<m), \tag{3}$$

and let $q_\mu>0$, p_μ be coprime pairs of integers such that

$$q_\mu^{r_\mu}\geqslant q_1^{r_1}\quad(1\leqslant\mu\leqslant m), \tag{4}$$
$$q_\mu^\omega\geqslant 2^{3m}\quad(1\leqslant\mu\leqslant m). \tag{5}$$

Suppose that $S(x_1,\dots,x_m)$ is a polynomial of degree at most r_μ in x_μ ($1\leqslant\mu\leqslant m$) with integer coefficients and

$$\overline{|S|}\leqslant q_1^{\omega r_1}, \tag{6}$$

but not vanishing identically. Then S has index at most ϵ at $(p_1/q_1,\dots,p_m/q_m)$ with respect to $(r_1,\dots,r_m)$.

Note. The precise form of (1)–(6) is immaterial. It is enough that the index in question is small (or, indeed, bounded above by an absolute constant) if the coefficients of S are not too big, the q_μ are large enough and the r_μ decrease quickly enough. That some such condition on the r_μ is necessary follows from the observation that $(x-y)^r$ has index at least 1 with respect to (r,r) at any $(p/q,p/q)$.

The proof is by induction and uses operators of the type

$$\Delta = \frac{\partial^{i_1+\dots+i_m}}{\partial x_1^{i_1} \dots \partial x_m^{i_m}}. \tag{7}$$

We call $i_1 + \dots + i_m$ the ORDER of Δ. If $\Delta_1, \dots, \Delta_h$ have order at most $0, \dots, h-1$ respectively and $\phi_1, \dots, \phi_h$ are functions of $x_1, \dots, x_m$ we call

$$\det(\Delta_i \phi_j) \quad (1 \leqslant i, j \leqslant h), \tag{8}$$

a (generalized) WRONSKIAN. When $m = 1$ there is precisely one Δ of order $i-1$, namely d^{i-1}/dx_1^{i-1}, so the one Wronskian which does not vanish trivially is the ordinary Wronskian

$$\det(d^{i-1} \phi_j / dx_1^{i-1}).$$

LEMMA 6. *Suppose that* $\phi_1, \dots, \phi_h$ *are rational functions†* of $x_1, \dots, x_m$ *and that*

$$c_1 \phi_1 + \dots + c_h \phi_h = 0, \tag{9}$$

for constants $c_1, \dots, c_h$ *implies* $c_1 = c_2 = \dots = c_h = 0$. *Then some Wronskian* (8) *does not vanish.*

Note. If (9) holds for some constants $c_1, \dots, c_h$ not all 0 it is trivial that conversely all the Wronskians vanish.

Proof. If $h = 1$ the only Wronskian is ϕ_1 itself, so the lemma is trivial. We shall therefore assume that $h > 1$ and that the lemma has already been proved for sets of fewer rational functions.

$\phi_1 = 0$ is a relation of type (9), so $\phi_1 \neq 0$. Write

$$\phi_j^* = \phi_1^{-1} \phi_j \quad (1 \leqslant j \leqslant h).$$

By the rules for differentiating a product every Wronskian of $\phi_1^*, \dots, \phi_h^*$ is expressible as a sum of Wronskians of $\phi_1, \dots, \phi_h$ multiplied by rational functions (products of derivatives of ϕ_1^{-1}). In particular, if $\phi_1^*, \dots, \phi_h^*$ have some non-vanishing Wronskian, so do $\phi_1, \dots, \phi_h$. Since any relation of type (9) for the ϕ_j^* implies one for the ϕ_j we may thus suppose without loss of generality that

$$\phi_1 = 1.$$

If now ϕ_h were a constant c, (say), there would be a relation $\phi_h - c\phi_1 = 0$ of type (9), contrary to hypothesis. Hence there is some variable, say x_1, such that

$$\partial \phi_h / \partial x_1 \neq 0. \tag{10}$$

† That is, quotients of polynomials.

On the other hand there may well be a linear combination

$$c_2\phi_2 + \ldots + c_h\phi_h, \qquad (11)$$

independent of x_1. If so, one of $c_2, \ldots, c_{h-1}$ is not 0, say $c_2 \neq 0$. Without loss of generality, $c_2 = 1$. We replace ϕ_2 by (11), which does not affect the Wronskians and gives

$$\partial\phi_2/\partial x_1 = 0.$$

By continuing in this way we may ultimately suppose that there is some k in $1 \leqslant k < h$, such that

$$\partial\phi_1/\partial x_1 = \ldots = \partial\phi_k/\partial x_1 = 0, \qquad (12)$$

but such that

$$e_{k+1}\,\partial\phi_{k+1}/\partial x_1 + \ldots + e_h\,\partial\phi_h/\partial x_1 = 0 \qquad (13)$$

for constants $e_{k+1}, \ldots, e_h$ implies $e_{k+1} = \ldots = e_h = 0$. By the hypothesis of the induction there are operators $\Delta_1^*, \ldots, \Delta_k^*$ of order at most $0, \ldots, k-1$ respectively such that

$$W_1 = \det(\Delta_i^*\phi_j) \neq 0 \quad (1 \leqslant i, j \leqslant k).$$

Similarly, since (13) has no non-trivial solution there are operators $\Delta_{k+1}^*, \ldots, \Delta_h^*$ of order at most $0, \ldots, h-k-1$ respectively such that

$$W_2 = \det(\Delta_i^*\,\partial\phi_j/\partial x_1) \neq 0 \quad (k < i, j \leqslant h).$$

Put

$$\Delta_i = \begin{cases} \Delta_i^* & (1 \leqslant i \leqslant k), \\ \Delta_i^* \dfrac{\partial}{\partial x_1} & (k < i \leqslant h), \end{cases}$$

so that Δ_i has order at most $i-1$. Then, by (12), we have

$$\det(\Delta_i\phi_j) = W_1 W_2 \neq 0 \quad (1 \leqslant i, j \leqslant h).$$

This proves the lemma.

COROLLARY. *If $\phi_1, \ldots, \phi_h$ have rational coefficients then it is enough to consider rational coefficients $c_1, \ldots, c_h$ in* (9).

Proof. For then only rational numbers occur in the proof. (Alternatively use Lemma 2 of Chapter III.)

Proof of Theorem IV $(m=1)$. If

$$S(p_1/q_1) = S'(p_1/q_1) = \ldots = S^{(t-1)}(p_1/q_1) = 0 \neq S^{(t)}(p_1/q_1),$$

then

$$S(x_1) = (x_1 - p_1/q_1)^t\, T(x_1)$$

for some polynomial $T(x_1)$. We have

$$S(x_1) = (q_1 x_1 - p_1)^t (q_1^{-t} T(x_1)),$$

where $q_1^{-t} T(x_1)$ has integer coefficients by Gauss's Lemma (Appendix C) since $S(x_1)$ has integer coefficients and

$$\text{g.c.d.} (p_1, q_1) = 1.$$

Hence the highest coefficient in $S(x_1)$ is divisible by q_1^t and *a fortiori*

$$q_1^t \leqslant \lceil S \rceil \leqslant q_1^{\omega r_1} = q_1^{er_1}$$

by (1) and (6). This proves the theorem for $m = 1$ since the index of S at p_1/q_1 with respect to r_1 is t/r_1 by definition.

Proof of Theorem IV $(m > 1)$. We use induction on m and so may assume the truth of the theorem for smaller values of m.

There certainly exist decompositions of the type

$$S = \sum_{1 \leqslant j \leqslant h} \phi_j(x_1, \ldots, x_{m-1}) \, \psi_j(x_m), \qquad (14)$$

where the ϕ_j, ψ_j are polynomials with rational (not necessarily integer) coefficients and the ϕ_j depend only on $x_1, \ldots, x_{m-1}$, the ψ_j only on x_m; for example with $h = r_m + 1$ and $\psi_j = x_m^{j-1}$. We take one such decomposition with the smallest possible h, so

$$h \leqslant r_m + 1. \qquad (14')$$

If there existed a linear relation $c_1 \phi_1 + \ldots + c_h \phi_h = 0$ with rational constants $c_1, \ldots, c_h$ and, say, $c_h \neq 0$, then

$$S = \sum_{1 \leqslant j < h} \phi_j(\psi_j - c_j \psi_h/c_h);$$

a decomposition into $h - 1$ summands. Since h is minimal, there can be no such relation. Similarly $c_1 \psi_1 + \ldots + c_h \psi_h = 0$ for rational constants $c_1, \ldots, c_h$ implies that $c_1 = \ldots = c_h = 0$. By Lemma 6, Corollary, we have

$$U(x_m) = \det \left(\frac{1}{(i-1)!} \frac{d^{i-1} \psi_j}{dx_m^{i-1}} \right) \neq 0 \quad (1 \leqslant i, j \leqslant h), \qquad (15)$$

where it is convenient to insert the numerical factors $((i-1)!)^{-1}$. By the same lemma there are Δ_i' $(1 \leqslant i \leqslant h)$ of the type

$$\Delta_i' = \frac{1}{i_1! \ldots i_{m-1}!} \frac{\partial^{i_1 + \ldots + i_{m-1}}}{\partial x_1^{i_1} \ldots \partial x_{m-1}^{i_{m-1}}} \qquad (16)$$

with $\qquad\qquad i_1 + \ldots + i_{m-1} \leqslant i - 1 \leqslant h - 1 \leqslant r_m,$ (17)

such that

$$V(x_1, \ldots, x_{m-1}) = \det(\Delta_i' \phi_j) \neq 0 \quad (1 \leqslant i, j \leqslant h).$$ (18)

We define

$$W(x_1, \ldots, x_m) = \det\left(\Delta_i' \frac{1}{(j-1)!} \frac{\partial^{j-1}}{\partial x_m^{j-1}} S(x_1, \ldots, x_m)\right)$$

$$(1 \leqslant i, j \leqslant h). \quad (19)$$

Then by (14), (15), (18) we have

$$W = \det\left(\Delta_i' \frac{1}{(j-1)!} \frac{\partial^{j-1}}{\partial x_m^{j-1}} \sum_k \phi_k \psi_k\right)$$

$$= U(x_m) V(x_1, \ldots, x_{m-1}).$$

But $\qquad\qquad \Delta_i' \dfrac{1}{(j-1)!} \dfrac{\partial^{j-1} S}{\partial x_m^{j-1}} = S_{i_1 \ldots i_{m-1}, j-1},$ (20)

if Δ_i' is given by (16); and so W has integer coefficients by (19) and Lemma 1. Hence

$$W(x_1, \ldots, x_m) = u(x_m) v(x_1, \ldots, x_{m-1})$$

for polynomials u, v with integer coefficients by Gauss's lemma (Appendix C).

Since (20) has degree at most r_μ in x_μ $(1 \leqslant \mu \leqslant m)$ the degree of the determinant W in x_μ is at most hr_μ; that is, $u(x_m)$ has degree at most hr_m and $v(x_1, \ldots, x_{m-1})$ has degree at most hr_μ in x_μ $(1 \leqslant \mu < m)$.

Now $\lceil S_{i_1 \ldots i_{m-1}, j-1} \rceil \leqslant 2^{r_1 + \ldots + r_m} q_1^{\omega r_1}$ by (6) and Lemma 1. Since there are at most $(r_1 + 1) \ldots (r_m + 1) \leqslant 2^{r_1 + \ldots + r_m}$ terms in any $S_{i_1 \ldots i_m}$ and since by (14') there are at most $h! \leqslant h^{h-1} \leqslant h^{rm} \leqslant 2^{hr_m}$ products in the expansion of W as a determinant, we have

$$\lceil W \rceil \leqslant h! ((r_1 + 1) \ldots (r_m + 1))^h \cdot (2^{r_1 + \ldots + r_m} q_1^{\omega r_1})^h$$

$$< (2^{3(r_1 + \ldots + r_m)} q_1^{\omega r_1})^h$$

$$\leqslant (2^{3m} q_1^{\omega})^{r_1 h}$$

$$\leqslant q_1^{2\omega r_1 h}$$

by (5), (19) and (20). Since u, v have integer coefficients, we have

$$\lceil u \rceil \leqslant q_1^{2\omega r_1 h}, \quad \lceil v \rceil \leqslant q_1^{2\omega r_1 h},$$ (21)

because every coefficient in $W = uv$ is the product of a coefficient in $u(x_m)$ by a coefficient in $v(x_1, \dots, x_{m-1})$.

Now we have

$$\omega = \omega(m, \epsilon) = \tfrac{1}{2}\omega(m - 1, \epsilon^2/12).$$

The theorem with $m - 1$ for m; $hr_1, \dots, hr_{m-1}$ for $r_1, \dots, r_{m-1}$; $\epsilon^2/12$ for ϵ, and so 2ω for ω thus applies to $v(x_1, \dots, x_{m-1})$: since (3) and (5) are stronger than the corresponding inequalities with 2ω instead of ω, and (21) replaces (6). Hence the index of v at $(p_1/q_1, \dots, p_{m-1}/q_{m-1})$ with respect to $(hr_1, \dots, hr_{m-1})$ is at most $\epsilon^2/12$. From the definition, the index of $v(x_1, \dots, x_{m-1})$ considered as a function of $x_1, \dots, x_m$ at $(p_1/q_1, \dots, p_m/q_m)$ with respect to $(r_1, \dots, r_m)$ is thus at most $h\epsilon^2/12$.

Similarly, since $q_1^{r_1} \leqslant q_m^{r_m}$ by (4) and since

$$\omega = \omega(m, \epsilon) \leqslant \tfrac{1}{2}\omega(1, \epsilon^2/12),$$

the theorem with 1 for m, hr_m for r_1, $\epsilon^2/12$ for ϵ applies to $u(x_m)$. Hence, as before, the index of $u(x_m)$ at $(p_1/q_1, \dots, p_m/q_m)$ with respect to $(r_1, \dots, r_m)$ is at most $h\epsilon^2/12$.

By Lemma 2 the index Θ of $W = uv$ at $(p_1/q_1, \dots, p_m/q_m)$ with respect to $(r_1, \dots, r_m)$ is

$$\Theta \leqslant \frac{h\epsilon^2}{12} + \frac{h\epsilon^2}{12} = \frac{h\epsilon^2}{6}. \tag{22}$$

We now estimate Θ in terms of θ, the index of $S(x_1, \dots, x_m)$ at $(p_1/q_1, \dots, p_m/q_m)$ with respect to $(r_1, \dots, r_m)$. By Lemma 2 (i) the corresponding index of $S_{i_1 \dots i_{m-1}, j-1}$ is at least

$$\theta - \frac{i_1}{r_1} - \dots - \frac{i_{m-1}}{r_{m-1}} - \frac{j-1}{r_m}$$

$$\geqslant \theta - \frac{i_1 + \dots + i_{m-1}}{r_{m-1}} - \frac{j-1}{r_m}$$

$$\geqslant \theta - \frac{r_m}{r_{m-1}} - \frac{j-1}{r_m} \quad \text{(by (17)}$$

$$\geqslant \theta - \omega - \frac{j-1}{r_m}$$

$$\geqslant \theta - \frac{\epsilon^2}{24} - \frac{j-1}{r_m} \quad (m > 1)$$

by (1), (3), (17). Since an index is always non-negative we deduce on developing the determinant (19) and using Lemma 2 that

$$\Theta \geqslant \sum_{1 \leqslant j \leqslant h} \max \left(\theta - \frac{\epsilon^2}{24} - \frac{j-1}{r_m}, 0 \right)$$

$$\geqslant -\frac{h\epsilon^2}{24} + \sum_{1 \leqslant j \leqslant h} \max \left(\theta - \frac{j-1}{r_m}, 0 \right).$$

Hence by (22) we have

$$h^{-1} \sum_{1 \leqslant j \leqslant h} \max \left(\theta - \frac{j-1}{r_m}, 0 \right) \leqslant \frac{\epsilon^2}{6} + \frac{\epsilon^2}{24} < \frac{\epsilon^2}{4}, \qquad (23)$$

where $1 \leqslant h \leqslant r_m + 1$.

If $\theta \geqslant (h-1)/r_m$ the left-hand side of (23) is

$$\tfrac{1}{2}\theta + \frac{1}{2} \left(\theta - \frac{h-1}{r_m} \right) \geqslant \tfrac{1}{2}\theta,$$

and so (23) gives $\theta < \tfrac{1}{2}\epsilon^2 < \epsilon$, the conclusion of the theorem.

Otherwise, $\theta < (h-1)/r_m$ and the left-hand side of (23) is

$$h^{-1} \sum_{0 \leqslant j-1 \leqslant \theta r_m} \left(\theta - \frac{j-1}{r_m} \right)$$

$$\geqslant h^{-1} ([\theta r_m] + 1) . \tfrac{1}{2}\theta$$

$$\geqslant \theta^2 r_m / 2h$$

$$\geqslant \theta^2 / 4,$$

since $h \leqslant r_m + 1 \leqslant 2r_m$. Hence (23) gives $\theta \leqslant \epsilon$, as required.

6. Proof of Theorem I. We shall suppose that

$$|\xi - p/q| < q^{-2-\delta}, \quad q > 0, \quad p, q \text{ coprime}, \qquad (1)$$

has infinitely many integer solutions. Theorem I will be proved if we can deduce a contradiction. There is no loss of generality in supposing that (4·2) holds, namely $0 < \delta < 1/12$.

We choose parameters successively as follows:

(i) ϵ is any number $< \delta/20$. Then (4·3) and (5·2) hold.

(ii) m is any integer $> 8n^2 \epsilon^{-2}$, that is, (3·1) holds; $\omega = \omega(m, \epsilon)$ is given by (5·1).

(iii) (p_1, q_1) is any solution of (1) with q_1 sufficiently large that (4·4), (5·5) hold for $\mu = 1$ and, moreover,

$$q_1^\omega > \gamma^m, \quad \gamma = 4(a+1). \qquad (2)$$

(iv) (p_μ, q_μ) are solutions of (1) chosen successively so that

$$\tfrac{1}{2}\omega \log q_{\mu+1} > \log q_\mu \quad (1 \leqslant \mu < m). \tag{3}$$

This can be done since (1) is assumed to have infinitely many solutions. Since $q_m > q_{m-1} > \ldots > q_1$ conditions (4·4), (5·5) hold for $1 \leqslant \mu \leqslant m$ by stage (iii).

(v) r_1 is any integer so large that

$$\epsilon r_1 \log q_1 \geqslant \log q_m. \tag{4}$$

(vi) For $2 \leqslant \mu \leqslant m$ put

$$r_\mu = \left[\frac{r_1 \log q_1}{\log q_\mu}\right] + 1. \tag{5}$$

Then, by (4),

$$r_1 \log q_1 \leqslant r_\mu \log q_\mu \leqslant r_1 \log q_1 + \log q_\mu$$
$$\leqslant (1 + \epsilon) r_1 \log q_1; \tag{6}$$

which is (4·5) and (5·4). Further, (3), (6) imply

$$\omega r_\mu \geqslant 2(1+\epsilon)^{-1} r_{\mu+1} \geqslant r_{\mu+1},$$

which is (5·3).

The conditions of Theorems II, III are satisfied. Further, the R given by Theorem II satisfies the conditions laid on S in Theorem IV since

$$\boxed{R} \leqslant \gamma^{r_1 + \ldots + r_m} < \gamma^{mr_1} < q_1^{\omega r_1}$$

by (3·3) and (2), and the other conditions of Theorem IV have been satisfied. Thus the index of R at $(p_1/q_1, \ldots, p_m/q_m)$ with respect to $(r_1, \ldots, r_m)$ is at least $\delta m/8$ by Theorem III and at most ϵ by Theorem IV. Hence $0 < \delta \leqslant 8\epsilon/m$. Since ϵ is arbitrarily small and m arbitrarily large, this is the required contradiction.

NOTES

See ROTH (1955) for a history of Theorem I. Earlier, much weaker, formulations are due to Thue, Siegel, Dyson, Gel'fond, Schneider, and these are usually cited as the 'Thue-Siegel Theorem'. For an explicit bound for the number of p, q in Theorem I see DAVENPORT & ROTH (1955). Theorem I enables the argument of §1 to be reversed and so gives a lower bound for $|q^n f(p/q)|$. This permits applications to certain Diophantine equations (cf. LANDAU (1927), **3**, 58–65).

Let $\nu > 1$. Theorem I states that $\| q\xi \| \geqslant q^{-\nu}$ for all q greater than some $q_0(\xi, \nu)$. It is at first sight paradoxical that there is no known way of finding an admissible $q_0(\xi, \nu)$ if $\nu < n - 1$.

Numbers which are not algebraic are TRANSCENDENTAL. Since the algebraic numbers are denumerable, 'almost all' numbers are transcendental in the sense of Chapter VII. Theorem I (or indeed Liouville's theorem of § 1) enables us to construct transcendental numbers: any number ξ such that

$$\| q\xi \| \leqslant q^{-\nu} \tag{*}$$

has infinitely many solutions for some $\nu > 1$ is transcendental, e.g. $\xi = \Sigma 2^{-3^n}$. [Put $q = 2^{3^n}$.] On the other hand, Theorem I of Chapter VII shows that (*) has only a finite number of solutions for almost all ξ; so this criterion of transcendence 'almost never' applies. There are proofs of the transcendence of e and π in HARDY & WRIGHT (1938). For the deep and rich theory of transcendence and related problems see SIEGEL (1949), GEL'FOND (1952) or SCHNEIDER (1956).[†] For the latest about rational approximation to e and π see MAHLER (1953 a, b).

† I have seen only a prospectus of Schneider's book.

METRICAL THEORY

1. Introduction. In only this chapter an acquaintance with
the rudiments of the theory of Lebesgue measure will be assumed.
As is usual, we shall say that ALMOST NO points of a certain
n-dimensional set have a certain property if the points of the
set which have the property have n-dimensional measure 0.
ALMOST ALL points of a set have a property if almost no points
of the set lack it. The measure of a set $\mathscr{E}$ will be denoted by $|\mathscr{E}|$.

In Chapter II we showed that the inequality

$$\| q\theta \| < C/q, \qquad (1)$$

has infinitely many integer solutions for all irrational θ if $C = 5^{-\frac{1}{2}}$.
This statement becomes false if C is given any value less than
$5^{-\frac{1}{2}}$; but, from our present point of view, this is merely an
accident due to the idiosyncrasies of the number $\frac{1}{2}(5^{\frac{1}{2}} - 1)$ and
the enumerably many θ associated with it. Otherwise a smaller
number than $5^{-\frac{1}{2}}$ will do; and indeed for any $C > \frac{1}{3}$ there are only
enumerably many exceptional θ for which (1) has only a finite
number of integer solutions q. If $C < \frac{1}{3}$, as we saw, the set of
exceptional θ is non-enumerable. However, as we shall now
show, (1) with any $C > 0$ has infinitely many solutions for almost
all θ; and indeed more is true:

THEOREM I. *Let $\psi(q)$ be a monotonely decreasing function of the
integer variable $q > 0$ with $0 \leqslant \psi(q) \leqslant \frac{1}{2}$. Then the set of inequalities*

$$\| q\theta_j \| < \psi(q) \quad (1 \leqslant j \leqslant n)$$

*has infinitely many integer solutions $q > 0$ for almost no or for
almost all sets of n numbers $(\theta_1, ..., \theta_n)$ according as*

$$\Sigma(\psi(q))^n$$

converges or diverges.

For example $\qquad \| q\theta \| < 1/q \log q$

has infinitely many solutions for almost all θ, which is a stronger
statement than that (1) has for any $C > 0$; but

$$\| q\theta \| < 1/q \log^2 q$$

has infinitely many solutions for almost no θ.

There is a complement to Theorem I for inhomogeneous approximation which is much easier to prove:

THEOREM II. *Let $0 \leqslant \psi(q) \leqslant \frac{1}{2}$ for all q. Then the set of inequalities*

$$\| q\theta_j - \alpha_j \| < \psi(q) \quad (1 \leqslant j \leqslant n),$$

has infinitely many integer solutions for almost no or for almost all $2n$-dimensional sets $(\theta_1, ..., \theta_n, \alpha_1, ..., \alpha_n)$ according as

$$\Sigma(\psi(q))^n$$

converges or diverges.

Note. $\psi(q)$ is not here required to be monotone.

It is clearly enough to consider only the θ_j, α_j with

$$0 \leqslant \theta_j < 1, \quad 0 \leqslant \alpha_j < 1.$$

We shall also for simplicity consider only $n = 1$, indicating the slight modifications required when $n > 1$ in § 7. All the sets occurring in the following arguments are easily seen to be measurable.

2. The convergence case ($n = 1$).

It is obvious that the convergence cases of both Theorems I and II follow at once from

LEMMA 1. *Suppose that $\Sigma\psi(q)$ is convergent, where $0 \leqslant \psi(q) \leqslant \frac{1}{2}$, and that α is fixed. Then*

$$\| q\theta - \alpha \| < \psi(q) \tag{1}$$

has infinitely many integer solutions $q > 0$ for almost no θ.

Proof. For fixed α, q the θ satisfying (1), i.e.

$$\left| \theta - \frac{p+\alpha}{q} \right| < \frac{\psi(q)}{q} \quad (p = \text{integer}),$$

form a set of intervals along the real axis of width $2\psi(q)/q$ with centres $1/q$ apart. The set of θ, $0 \leqslant \theta < 1$, satisfying (1) thus has measure $2\psi(q)$. Hence the set of θ for which (1) has a solution with $q \geqslant Q$ has measure at most

$$2 \sum_{q \geqslant Q} \psi(q) < \epsilon$$

122 DIOPHANTINE APPROXIMATION

for any $\epsilon > 0$, if Q is large enough. In particular, the set of θ for which (1) has infinitely many solutions has measure at most ϵ. Since ϵ is arbitrary, this proves the lemma.

3. Two lemmas. Let $f(x,y) \geqslant 0$, $g(x,y) \geqslant 0$ be defined in, say, the unit square
$$\mathscr{I}: 0 \leqslant x < 1, \quad 0 \leqslant y < 1. \tag{1}$$
Then the well-known inequality of Schwarz states that†

$$\left(\iint_{\mathscr{I}} fg\,dx\,dy\right)^2 \leqslant \left(\iint_{\mathscr{I}} f^2\,dx\,dy\right)\left(\iint_{\mathscr{I}} g^2\,dx\,dy\right). \tag{2}$$

In particular $(g=1)$,

$$M_1 = \iint_{\mathscr{I}} f\,dx\,dy \leqslant M_2 = \left(\iint_{\mathscr{I}} f^2\,dx\,dy\right)^{\frac{1}{2}}.$$

LEMMA 2. (Paley-Zygmund.) *Suppose that* $f(x,y) \geqslant 0$, *that* $M_1 \geqslant aM_2$, *and that* $0 \leqslant b \leqslant a$. *Then the set* $\mathscr{E}$ *in which*
$$f(x,y) \geqslant bM_2 \ (\geqslant bM_1)$$
has measure $|\mathscr{E}| \geqslant (a-b)^2$.

Note. There are analogous results for functions of any number of variables.

Proof.
$$\left(\iint_{\mathscr{E}} f\,dx\,dy\right)^2 \leqslant \left(\iint_{\mathscr{E}} dx\,dy\right)\left(\iint_{\mathscr{E}} f^2\,dx\,dy\right)$$
$$\leqslant |\mathscr{E}| \iint_{\mathscr{I}} f^2\,dx\,dy = |\mathscr{E}| M_2^2 \tag{3}$$

by Schwarz's inequality. As $f \leqslant bM_2$ in‡ $\mathscr{I} - \mathscr{E}$ we have

$$\iint_{\mathscr{E}} f\,dx\,dy = \iint_{\mathscr{I}} f\,dx\,dy - \iint_{\mathscr{I}-\mathscr{E}} f\,dx\,dy$$
$$\geqslant M_1 - bM_2$$
$$\geqslant (a-b)M_2. \tag{4}$$

The lemma now follows at once from (3) and (4).

---

† This is the integral analogue of $(\Sigma a_j b_j)^2 \leqslant (\Sigma a_j^2)(\Sigma b_j^2)$. Perhaps the simplest proof of (2) is to note that the quadratic form $h(X, Y) = \iint (Xf+Yg)^2\,dx\,dy \geqslant 0$ and that the difference between the two sides of (2) is the discriminant of h.
‡ $\mathscr{I} - \mathscr{E}$ is the set of points belonging to $\mathscr{I}$ but not to $\mathscr{E}$.

LEMMA 3. *Let $\delta(x)$ be a function of period 1 of the real variable x.*
Then

$$\int_0^1 \delta(qx+\alpha)\,dx = \int_0^1 \delta(x)\,dx,$$

for any real α and integer $q \neq 0$.
  Proof.

$$\int_0^1 \delta(qx+\alpha)\,dx = \int_{\alpha/q}^{1+\alpha/q} \delta(qx)\,dx = \int_0^1 \delta(qx)\,dx.$$

$$= \frac{1}{q}\int_0^q \delta(y)\,dy = \int_0^1 \delta(x)\,dx,$$

since, e.g. for $q > 0$,

$$\int_0^q \delta(y)\,dy = \int_0^1 + \int_1^2 + \ldots + \int_{q-1}^q.$$

4. Proof of Theorem II (divergence, $n = 1$). Let $\Delta_Q(\theta,\alpha)$ be the number of integer solutions of

$$\| q\theta - \alpha \| < \psi(q), \quad 0 < q \leqslant Q.$$

We shall apply Lemma 2 to $\Delta_Q(\theta,\alpha)$ and write

$$M_1(Q) = \iint \Delta_Q(\theta,\alpha)\,d\theta\,d\alpha,$$

$$M_2(Q) = \left(\iint \Delta_Q^2(\theta,\alpha)\,d\theta\,d\alpha \right)^{\frac{1}{2}}.$$

Here, unless the contrary is explicitly stated, all integrals are over the unit square

$$0 \leqslant \theta < 1, \quad 0 \leqslant \alpha < 1.$$

To estimate $M_1(Q)$ and $M_2(Q)$ put

$$\delta_q(x) = \begin{cases} 1 & \text{if} \quad \| x \| < \psi(q) \\ 0 & \text{otherwise,} \end{cases}$$

so that

$$\Delta_Q(\theta,\alpha) = \sum_{q \leqslant Q} \delta_q(q\theta - \alpha).$$

The sum $\qquad\qquad \Psi(Q) = \sum\limits_{q \leqslant Q} \psi(q) \to \infty$

by the assumption of divergence.

LEMMA 4.

(i) $\qquad\qquad\qquad \iint \delta_q(q\theta - \alpha)\, d\theta\, d\alpha = 2\psi(q).$

(ii) $\quad \iint \delta_q(q\theta - \alpha)\, \delta_r(r\theta - \alpha)\, d\theta\, d\alpha = \begin{cases} 4\psi(q)\,\psi(r) & (q \neq r), \\ 2\psi(q) & (q = r). \end{cases}$

Proof (i). Trivial by Lemma 3 and since clearly

$$\int_0^1 \delta_q(x)\, dx = 2\psi(q).$$

Proof (ii). The left-hand side of (ii) is

$$\iint \delta_q(-\alpha')\, \delta_r(s\theta - \alpha')\, d\theta\, d\alpha',$$

where $s = r - q$, $\alpha' = \alpha - q\theta$; and the range of α' can be taken as $0 \leqslant \alpha' < 1$ since δ_q, δ_r are periodic. If $r \neq q$, so $s \neq 0$, we have

$$\int_0^1 \delta_r(s\theta - \alpha')\, d\theta = \int_0^1 \delta_r(x)\, dx = 2\psi(r)$$

by Lemma 3, and (ii) follows on integrating again. However, if $r = q$, the integrand is $\delta_q^2(-\alpha') = \delta_q(-\alpha')$ since $\delta_q(x) = 0$ or 1; and again the result is immediate.

COROLLARY. *Let $\epsilon > 0$ be arbitrarily small. Then*

$$2\Psi(Q) = M_1(Q) \geqslant (1 - \epsilon)\, M_2(Q)$$

for all sufficiently large Q.

Proof. First,

$$M_1(Q) = \iint \Delta_Q\, d\theta\, d\alpha = \sum_{q \leqslant Q} \iint \delta_q(q\theta - \alpha)\, d\theta\, d\alpha$$

$$= 2 \sum_{q \leqslant Q} \psi(q) = 2\Psi(Q).$$

Secondly,

$$M_2^2(Q) = \iint \Delta_Q^2 \, d\theta \, d\alpha$$

$$= \sum_{q,\, r \leqslant Q} \iint \delta_q(q\theta - \alpha)\, \delta_r(r\theta - \alpha)\, d\theta \, d\alpha$$

$$= 2 \sum_{q \leqslant Q} \psi(q) + 4 \sum_{\substack{q,\, r \leqslant Q \\ q \neq r}} \psi(q)\, \psi(r)$$

$$\leqslant 2\Psi(Q) + 4\Psi^2(Q)$$

$$\leqslant (1-\epsilon)^{-2}\, 4\Psi^2(Q)$$

for all large enough Q, since $\Psi(Q) \to \infty$.

Proof of Theorem II (*resumed*). By Lemma 2 with $a = 1 - \epsilon$, $b = \epsilon$ we have
$$\Delta_Q(\theta, \alpha) \geqslant \epsilon M_1(Q) = 2\epsilon\Psi(Q),$$

in a set in the unit square of measure at least $(1 - 2\epsilon)^2 \geqslant 1 - 4\epsilon$. Since $\Delta_Q(\theta, \alpha)$ increases monotonely with Q it follows that $\Delta_Q(\theta, \alpha) \to \infty$ as $Q \to \infty$ in the unit square except, possibly, in a set of measure 4ϵ. Since ϵ is arbitrarily small, this is just the assertion of the theorem.

5. Some more lemmas.

LEMMA 5. *Let $\mathscr{E}$ be a set of measure $|\mathscr{E}| > 0$ in the interval $0 \leqslant x < 1$ and let $\epsilon > 0$ be arbitrarily small. Then there are integers t, T with $0 \leqslant t < T$ such that the part of $\mathscr{E}$ in*

$$t/T \leqslant x < (t+1)/T, \tag{1}$$

has measure at least $(1 - \epsilon)/T$.

Proof. By the definition of measure there is a finite or enumerable set of non-overlapping intervals $\mathscr{I}_r$ which cover $\mathscr{E}$ and such that $\Sigma |\mathscr{I}_r| < (1 - \tfrac{1}{2}\epsilon)^{-1} |\mathscr{E}|$. But† $|\mathscr{E}| = \Sigma |\mathscr{I}_r \cap \mathscr{E}|$, and so

$$|\mathscr{I}_r \cap \mathscr{E}| > (1 - \tfrac{1}{2}\epsilon) |\mathscr{I}_r| \tag{2}$$

for at least one r.

We may choose now an interval $\mathscr{K}$: $x_0 \leqslant x < x_1$ with rational end-points x_0, x_1 which encloses this $\mathscr{I}_r$ and such that

$$|\mathscr{K}| < (1 - \tfrac{1}{2}\epsilon)^{-1} |\mathscr{I}_r|. \tag{3}$$

† $\mathscr{A} \cap \mathscr{B}$ is the set of points common to $\mathscr{A}$ and $\mathscr{B}$.

Then

$$|\mathcal{K} \cap \mathscr{E}| \geqslant |\mathscr{S}_r \cap \mathscr{E}| > (1 - \tfrac{1}{2}\epsilon)^2 |\mathcal{K}| > (1-\epsilon)|\mathcal{K}|$$

by (2) and (3).

Now let T be an integer such that $Tx_0 = t_0$, $Tx_1 = t_1$ are integers and denote the interval (1) by $\mathscr{L}_t$. Clearly

$$\sum_{t_0 \leqslant t < t_1} |\mathscr{L}_t \cap \mathscr{E}| = |\mathcal{K} \cap \mathscr{E}| > (1-\epsilon)|\mathcal{K}| = (1-\epsilon)(t_1 - t_0)/T,$$

and so $|\mathscr{L}_t \cap \mathscr{E}| > (1-\epsilon)/T$ for at least one t, as asserted.

COROLLARY. *Almost all numbers θ_1 have the form*

$$\theta_1 \equiv T\theta \quad (\mathrm{mod}\, 1), \quad T \ a \ positive \ integer; \quad \theta \in \mathscr{E}. \tag{4}$$

Proof. Let $\mathscr{E}_1$ be the set of θ_1 in $0 \leqslant \theta < 1$ of the form (4) and let $\epsilon > 0$ be arbitrarily small. If t, T are as in the lemma, the points

$$\theta_1 = T\theta - t, \quad t/T \leqslant \theta < (t+1)/T, \quad \theta \in \mathscr{E},$$

are in $\mathscr{E}_1$ and have measure $> 1 - \epsilon$. Hence $|\mathscr{E}_1| > 1 - \epsilon$. Since ϵ is arbitrary, this proves the corollary.

LEMMA 6. *Let $\phi(q)$ be the number of integers p in $0 < p < q$ which are prime to q. Then*

$$\sum_{q \leqslant Q} q^{-1}\phi(q) \geqslant C_1 Q$$

for all $Q > 1$ and some constant $C_1 > 0$.

Note. Roughly speaking this means that $q^{-1}\phi(q)$ is greater than C_1 'on the average'.

Proof. It is well known (e.g. HARDY & WRIGHT (1938)) that

$$Q^{-2}\Phi(Q) \to 3/\pi^2; \quad \Phi(Q) = \sum_{q \leqslant Q} \phi(q).$$

Hence $Q^{-2}\Phi(Q) \geqslant C_1 > 0$ for all $Q > 1$ and for some $C_1 > 0$, since $\Phi(Q) > 0$ if $Q > 1$. Then, using the device of 'partial summation', we have

$$\begin{aligned}
\sum_{q \leqslant Q} q^{-1}\phi(q) &= \sum_{q \leqslant Q} q^{-1}(\Phi(q) - \Phi(q-1)) \\
&= \sum_{q < Q} \Phi(q)(q^{-1} - (q+1)^{-1}) + Q^{-1}\Phi(Q) \\
&\geqslant Q^{-1}\Phi(Q) \\
&\geqslant C_1 Q,
\end{aligned}$$

as asserted.

[Alternatively $q^{-1} \phi(q) = \Sigma d^{-1} \mu(d)$ where the summation is over all divisors $d > 0$ of q and $\mu(d)$ is Möbius's function. Hence $Q^{-1} \sum_{q \leqslant Q} q^{-1} \phi(q) = \Sigma \mu(d) [d^{-1} Q]/dQ$ where now d runs through all integers. Since Σd^{-2} is convergent the right-hand side is uniformly convergent and tends to $\Sigma d^{-2} \mu(d) = 6\pi^{-2}$ as $Q \to \infty$.]

LEMMA 7. *Let $\omega(q)$ be monotonely decreasing and positive. Then*

$$\sum_{q \leqslant Q} q^{-1} \phi(q) \omega(q) \geqslant C_1 \sum_{1 < q \leqslant Q} \omega(q).$$

Proof. Write $\chi(Q) = \sum_{q \leqslant Q} q^{-1} \phi(q)$. Then, as in the previous proof,

$$\sum_{q \leqslant Q} q^{-1} \phi(q) \omega(q) = \sum_{q \leqslant Q} \omega(q) (\chi(q) - \chi(q-1))$$

$$= \sum_{q < Q} \chi(q) (\omega(q) - \omega(q+1)) + \chi(Q) \omega(Q)$$

$$\geqslant \sum_{1 < q < Q} C_1 q(\omega(q) - \omega(q+1)) + C_1 Q \omega(Q)$$

$$= C_1 \omega(2) + C_1 \sum_{1 < q \leqslant Q} \omega(q).$$

LEMMA 8. *Suppose that $f(x) \geqslant 0$ and that $f(x)$ increases for $x \leqslant 0$ and decreases for $x \geqslant 0$ as x increases. Then*

$$\sum_{\substack{q \neq 0 \\ -\infty < q < \infty}} f(q) \leqslant \int_{-\infty}^{\infty} f(x) \, dx,$$

if the right-hand side exists.

Proof. Clear.

6. Proof of Theorem I (divergence, $n = 1$).

Since $\Sigma \psi(q)$ diverges we may find $\tau(q)$, $0 < \tau(q) \leqslant 1$, decreasing monotonely to 0 but so slowly that†

$$\Omega(Q) = \sum_{q \leqslant Q} \omega(q) \to \infty,$$

where

$$\omega(q) = \tau(q) \psi(q).$$

Then $0 < \omega(q) \leqslant \frac{1}{2}$ since $0 < \psi(q) \leqslant \frac{1}{2}$ and $\omega(q)$ tends to 0 monotonely since $\psi(q)$ is monotone. We operate at first with $\omega(q)$ instead of $\psi(q)$.

Let

$$\beta_q(\theta) = \begin{cases} 1 & \text{if} \quad |\theta| < q^{-1} \omega(q), \\ 0 & \text{otherwise,} \end{cases}$$

---

† For there exist $1 = q_1 < q_2 < q_3 < \dots$ such that $\sum_{q_j \leqslant q < q_{j+1}} \psi(q) > 1$. Put $\tau(q) = j^{-1}$ if $q_j \leqslant q < q_{j+1}$.

where q is a positive integer and θ is real. Put

$$\gamma_q(\theta) = \sum_p' \beta_q(\theta - p/q),$$

where, as throughout the rest of the chapter, $\sum_p'$ is taken over

$$0 < p < q, \quad p \text{ prime to } q. \tag{1}$$

Then $\gamma_q(\theta)$ is the number of solutions of $|q\theta - p| < \omega(q)$ with p subject to (1); so

$$\gamma_q(\theta) = 0 \quad \text{or} \quad 1, \tag{2}$$

since $\omega(q) \leqslant \tfrac{1}{2}$. We shall apply the 1-dimensional analogue of Lemma 2 to

$$\Gamma_Q(\theta) = \sum_{q \leqslant Q} \gamma_q(\theta),$$

and write

$$M_1(Q) = \int_0^1 \Gamma_Q(\theta)\,d\theta, \quad M_2^2(Q) = \int_0^1 \Gamma_Q^2(\theta)\,d\theta. \tag{3}$$

LEMMA 9. *There is an absolute constant $C_2 > 0$ such that*

$$M_1(Q) \geqslant C_2\,\Omega(Q)$$

if Q is large enough.

   Proof.

$$\int_0^1 \gamma_q(\theta)\,d\theta = \int_0^1 \sum_p' \beta_q(\theta - p/q)\,d\theta$$

$$= \phi(q) \int_{-\infty}^\infty \beta_q(\theta)\,d\theta$$

$$= 2q^{-1}\phi(q)\,\omega(q). \tag{3'}$$

Hence, by Lemma 7,

$$M_1(Q) = 2 \sum_{q \leqslant Q} q^{-1}\phi(q)\,\omega(q) \geqslant 2C_1(\Omega(Q) - \omega(1))$$

$$\geqslant C_2\,\Omega(Q)$$

for any $C_2 < 2C_1$ if Q is large enough, since $\Omega(Q) \to \infty$.

   LEMMA 10.

$$\int_0^1 \gamma_q(\theta)\,\gamma_r(\theta)\,d\theta \leqslant 4\omega(q)\,\omega(r) \quad (q \neq r).$$

   Proof. Write

$$\lambda_{qr}(x) = \int_{-\infty}^\infty \beta_q(\theta)\,\beta_r(\theta - x)\,d\theta, \tag{4}$$

Clearly $\displaystyle\int_{-\infty}^{\infty}\lambda_{qr}(x)\,dx = \int_{-\infty}^{\infty}\int_{-\infty}^{\infty}\beta_q(\theta)\,\beta_r\,(\theta-x)\,d\theta\,dx$

$$= \left(\int_{-\infty}^{\infty}\beta_q(\theta)\,d\theta\right)\left(\int_{-\infty}^{\infty}\beta_r(y)\,dy\right)$$

$$= (2q^{-1}\,\omega(q))\,(2r^{-1}\omega(r)), \tag{5}$$

on putting $y=\theta-x$. Further, the integrand in (4) is 1 if both $|\theta| < q^{-1}\,\omega(q),\ |\theta-x| < r^{-1}\,\omega(r)$ and is otherwise 0. Hence† $\lambda_{qr}(x)$ decreases for $x\geqslant 0$ and increases for $x\leqslant 0$. But now, if $\sum\limits_s'$ is taken over (1) with (s,r) for (p,q),

$$\int_0^1 \gamma_q(\theta)\,\gamma_r(\theta)\,d\theta = \sum_p{}'\sum_s{}'\int_0^1 \beta_q(\theta-p/q)\,\beta_r(\theta-s/r)\,d\theta$$

$$\leqslant \sum_{\substack{0<p<q\\0<s<r\\p/q\neq s/r}} \lambda_{qr}\left(\frac{s}{r}-\frac{p}{q}\right) \tag{6}$$

on writing θ for $\theta-p/q$, and since $p/q \neq s/r$ if p is prime to q and s to r, $q \neq r$. Here

$$s/r - p/q = (qs-pr)/qr = ck/qr, \quad c = \text{g.c.d.}\,(q,r),$$

where $k \neq 0$ is an integer. Further, each k can occur at most c times, since if $q = cq_1,\ r = cr_1$ we have $q_1 s - r_1 p = k$, which determines p modulo q_1. Thus (6) is

$$\leqslant c \sum_{k=\pm 1,\,\pm 2,\,\ldots} \lambda_{qr}(ck/qr)$$

$$\leqslant c \int_{-\infty}^{\infty} \lambda_{qr}(cx/qr)\,dx$$

$$= 4\omega(q)\,\omega(r)$$

by (5) and Lemma 8.

LEMMA 11. $M_2^2(Q) \leqslant 5\Omega^2(Q)$

if Q is large enough.

   Proof.

$$M_2^2(Q) = \int_0^1 (\sum_{q\leqslant Q}\gamma_q(\theta))^2\,d\theta$$

$$= \sum_{q\leqslant Q}\int_0^1 \gamma_q^2(\theta)\,d\theta + \sum_{\substack{q,\,r\leqslant Q\\q\neq r}}\int_0^1 \gamma_q(\theta)\,\gamma_r(\theta)\,d\theta. \tag{7}$$

† It is, of course, easy to evaluate $\lambda_{qr}(x)$ explicitly.

But by (2) and (3′) the first sum is

$$\Sigma \int_0^1 \gamma_q(\theta)\, d\theta = 2\Sigma\, q^{-1}\, \phi(q)\, \omega(q) \leqslant 2\Sigma \omega(q)$$
$$= 2\Omega(Q).$$

The second sum, by Lemma 10, is

$$\leqslant 4 \sum_{q,\, r \leqslant Q} \omega(q)\, \omega(r) = 4\Omega^2(Q).$$

Hence $\qquad M_2^2(Q) \leqslant 2\Omega(Q) + 4\Omega^2(Q) \leqslant 5\Omega^2(Q)$

if Q is large enough, since $\Omega(Q) \to \infty$.

LEMMA 12. $\Gamma_Q(\theta) \to \infty$ *in a set $\mathscr{E}$ in $0 \leqslant \theta < 1$ of measure* $|\mathscr{E}| > 0$.

Proof. From Lemmas 9, 11 we see that

$$M_1(Q) \geqslant C_3 M_2(Q)$$

for some $C_3 > 0$ and all large enough Q. Hence

$$\Gamma_Q(\theta) \geqslant \tfrac{1}{2} C_3 M_1(Q) \geqslant \tfrac{1}{2} C_3 C_2\, \Omega(Q),$$

in a set of measure at least $(\tfrac{1}{2}C_3)^2$ by Lemmas 2 and 9. Since $\Gamma_Q(\theta)$ is monotone in Q and $\Omega(Q) \to \infty$ the lemma follows with $|\mathscr{E}| \geqslant (\tfrac{1}{2}C_3)^2$.

Proof of Theorem I (concluded). $\Gamma_Q(\theta)$ is the number of solutions of $|q\theta - p| < \omega(q)$ with $0 < q \leqslant Q$ and p subject to (1). Lemma 12 thus shows that

$$\| q\theta \| < \omega(q) \qquad (8)$$

has infinitely many integer solutions q in a set $\mathscr{E}$ in $0 \leqslant \theta < 1$ with $|\mathscr{E}| > 0$. Let θ_1 be any number such that

$$\theta_1 \equiv T\theta \pmod 1, \quad \theta \in \mathscr{E} \qquad (9)$$

for some integer $T > 0$. Now (8) implies

$$\| q\theta_1 \| = \| Tq\theta \| < T\omega(q)$$
$$= T\tau(q)\, \psi(q)$$
$$< \psi(q)$$

if q is large enough, since $\tau(q) \to 0$. Hence $\| q\theta_1 \| < \psi(q)$ has infinitely many solutions. But almost every number is of the form θ_1 in (9) by Lemma 5, Corollary.

7. The case $n \geqslant 2$. The modifications in the proof of Theorem II and in the convergence case of Theorem I are quite trivial; so it remains to consider Theorem I when $\Sigma \psi^n(q)$ diverges. We choose a monotone decreasing function $\tau(q)$ so that this time $\omega(q) = \tau(q)\,\psi(q)$ makes $\Sigma \omega^n(q)$ diverge. We define $\beta_q(\theta)$, $\gamma_q(\theta)$ as before and define $M_1(Q)$, $M_2(Q)$ in terms of

$$\Gamma_Q(\theta_1, \ldots, \theta_n) = \sum_{q \leqslant Q} \gamma_q(\theta_1) \ldots \gamma_q(\theta_n)$$

instead of $\Gamma_Q(\theta)$. The only modification of any depth required in the proof is in the analogue of Lemma 9 since

$$M_1(Q) = \sum_{q \leqslant Q} q^{-n}\, \phi^n(q)\, \omega^n(q).$$

But by Lemma 6 and the well-known† inequality

$$(Q^{-1} \sum_{q \leqslant Q} x_q^r)^{1/r} \leqslant (Q^{-1} \sum_{q \leqslant Q} x_q^s)^{1/s}$$

valid for all r, s with $0 < r \leqslant s$ and all positive x_q we have

$$Q^{-1} \sum_{q \leqslant Q} q^{-n}\, \phi^n(q) \geqslant (Q^{-1} \sum_{q \leqslant Q} q^{-1} \phi(q))^n \geqslant C_1^n$$

for all $Q > 1$.

NOTES

§ 1. It is a common feature of results of this type that there is no half-way house between 'almost all' and 'almost none'. Theorem I becomes false if $\psi(q)$ is not assumed to be monotone. For a discussion of these points see CASSELS (1950a).

§ 5. Lemma 5 is, of course, a consequence of the fact that a measurable set has density 1 at almost all its points.

Of course the 'metrical' approach can be made to most problems. Thus the discrepancy D_Q modulo 1 (in the sense of Chapter IV) of the sequence $q\theta$ is $O(Q^{-1}\log^{1+\epsilon}Q)$ for any $\epsilon > 0$ and almost all θ (KHINTCHINE (1923)). Much work has been done on the metrical theory of the uniform distribution of sequences of the type $f(q, \theta)$ (e.g. $f = q^r\theta$, r fixed) but the theory is in an unsatisfactory state (e.g. CASSELS (1950b, c)). Again, much is known of the behaviour of the partial quotients a_n of θ for almost all θ (KOKSMA (1936), Kap. III, § 29; KHINTCHINE (1935)).

---

† See, for example, HARDY, LITTLEWOOD & PÓLYA (1934), Theorem 16. The inequality is, of course, an immediate consequence of Hölder's.

Instead of considering independent $\theta_1, \ldots, \theta_n$ we may consider powers $\theta_j = \theta^j$ ($1 \leqslant j \leqslant n$). Mahler made the interesting conjecture that $\max \| q\theta^j \| \leqslant q^{-(1/n)-\epsilon}$ has only a finite number of integer solutions q for any $\epsilon > 0$ and almost all θ. Since the set of $(\theta, \ldots, \theta^n)$ has n-dimensional measure 0, Theorem I, while suggestive, is uninformative. For latest results see CASSELS (1951) and LEVEQUE (1953).

Again, if $\Sigma \psi_j(q)$ ($j = 1, 2$) both converge the sets $\mathscr{E}_j$ of θ for which $\| q\theta \| < \psi_j(q)$ ($j = 1, 2$) respectively have infinitely many solutions, while both of measure 0, may yet be of quite different 'fractional dimension' (cf. KOKSMA (1936), Kap. V, § 12).

CHAPTER VIII

THE PISOT-VIJAYARAGHAVAN NUMBERS

1. Introduction. In only this chapter a slight knowledge of the elements of the theory of algebraic numbers is assumed.

The PISOT-VIJAYARAGHAVAN NUMBERS (PV-NUMBERS) are those algebraic integers $\alpha > 1$ all of whose conjugates, other than α itself, lie in the open unit circle $|z| < 1$. In particular if $\alpha > 1$ is a rational integer it has no other conjugates and so is a PV-number. Let α be a PV-number of degree $r \geqslant 1$ with conjugates† $\alpha = \alpha_1, \alpha_2, \ldots, \alpha_r$ so that

$$\alpha = \alpha_1 > 1, \quad |\alpha_j| < 1 \quad (j \neq 1).$$

The trace (spur)

$$T(\alpha^n) = \alpha_1^n + \ldots + \alpha_r^n = A_n$$

is a rational integer for all integers $n \geqslant 0$, and so

$$\| \alpha^n \| \leqslant | \alpha^n - A_n | \leqslant | \alpha_2 |^n + \ldots + | \alpha_r |^n \to 0 \quad (n \to \infty).$$

We shall show that the PV-numbers are characterized by this property.

More generally let the irreducible equation for the PV-number α be

$$\alpha^r + a_{r-1}\alpha^{r-1} + \ldots + a_0 = 0, \tag{1}$$

where $a_0, \ldots, a_{r-1}$ are rational integers and suppose that λ is a number of the field of α such that

$$T(\lambda\alpha^N), \ T(\lambda\alpha^{N+1}), \ \ldots, \ T(\lambda\alpha^{N+r-1})$$

are all integers for some integer $N \geqslant 0$. For any integer $n \geqslant N + r$ we have, by (1),

$$0 = T(\lambda\alpha^{n-r} (\alpha^r + a_{r-1}\alpha^{r-1} + \ldots + a_0))$$
$$= T(\lambda\alpha^n) + a_{r-1} T(\lambda\alpha^{n-1}) + \ldots + a_0 T(\lambda\alpha^{n-r});$$

and so, by induction,

$$T(\lambda\alpha^n) = \text{integer} \quad (\text{all } n \geqslant N).$$

† Of course α_j is not necessarily real for $j \neq 1$.

Hence, as before,

$$\| \lambda \alpha^n \| \leqslant | \lambda_2 \alpha_2^n | + \ldots + | \lambda_r \alpha_r^n | \quad (n \geqslant N) \tag{2}$$
$$\to 0 \quad (n \to \infty),$$

where $\lambda_2, \ldots, \lambda_r$ are the conjugates of $\lambda = \lambda_1$. We shall prove the following converses:

THEOREM I. (Pisot, Vijayaraghavan.) *Suppose that $\alpha > 1$ is an algebraic number, that $\lambda \neq 0$ is real, and that*

$$\| \lambda \alpha^n \| \to 0 \quad (n \to \infty). \tag{3}$$

Then α is a PV-number: and $\lambda = \alpha^{-N} \mu$ for some integer $N \geqslant 0$ and for some number μ of the field of α such that $T(\alpha^j \mu) = integer$ $(0 \leqslant j \leqslant r - 1)$, where r is the degree of α.

THEOREM II. (Pisot.) *Suppose that $\alpha > 1$ and $\lambda \neq 0$ are real, and that*

$$\sum_{0 \leqslant n < \infty} \| \lambda \alpha^n \|^2 < \infty. \tag{4}$$

Then α is algebraic and the conclusions of Theorem I apply.

Of course (4) is much stronger than (3) but it is clear from (2) that conversely both (3) and (4) hold if α is a PV-number and λ is as specified in Theorem I. Whether Theorem II would remain true if (3) were substituted for (4) is an open question. From Theorem II we shall deduce the remarkable

THEOREM III. (Salem.) *The set of all PV-numbers is closed.*†

2. Proof of Theorem I. Throughout this section let the algebraic number α of degree r with conjugates $\alpha = \alpha_1, \alpha_2, \ldots, \alpha_r$ satisfy the irreducible equation

$$f(\alpha) = 0, \quad f(x) = a_r x^r + \ldots + a_0, \tag{1}$$

where $a_r, \ldots, a_0$ are integers.

LEMMA 1. *The set of r equations*

$$y_j = x_1 \alpha_1^j + \ldots + x_r \alpha_r^j \quad (0 \leqslant j < r), \tag{2}$$

† That is, if $\alpha^{(n)}$ $(n = 1, 2, \ldots)$ are all PV-numbers and $\alpha = \lim \alpha^{(n)}$ then α is also a PV-number.

has the unique solution

$$\delta_j x_j = \sum_{0 \leqslant k < r} \beta_{jk} y_k \quad (1 \leqslant j \leqslant r), \tag{3}$$

where

$$\delta_j = \sum_{1 \leqslant l \leqslant r} l a_l \alpha_j^{l-1} \neq 0, \quad \beta_{jk} = \sum_{k < l \leqslant r} a_l \alpha_j^{l-k-1}. \tag{3'}$$

Note. The exact form of δ_j, β_{jk} is immaterial. It is enough that they are polynomials in α_j with rational integer coefficients and that $\delta_j \neq 0$.

Proof. Since the α_j are distinct a unique solution $x_1, \ldots, x_r$ exists. The polynomial

$$f_j(z) = a_r \prod_{k \neq j} (z - \alpha_k) \tag{4}$$

is given by

$$f_j(z) = \frac{f(z) - f(\alpha_j)}{z - \alpha_j}$$

$$= \sum_{0 \leqslant l \leqslant r} a_l \frac{z^l - \alpha_j^l}{z - \alpha_j}$$

$$= \sum_{1 \leqslant l \leqslant r} a_l (z^{l-1} + \alpha_j z^{l-2} + \ldots + \alpha_j^{l-1})$$

$$= \sum_{0 \leqslant k < r} \beta_{jk} z^k.$$

Hence, by (2), (4) we have

$$\sum_k \beta_{jk} y_k = f_j(\alpha_1) x_1 + \ldots + f_j(\alpha_r) x_r = \delta_j x_j,$$

where

$$\delta_j = f_j(\alpha_j) = a_r \prod_{k \neq j} (\alpha_j - \alpha_k) \neq 0.$$

LEMMA 2. *Suppose that* $A_1, A_2, \ldots$ *are numbers and that*

$$a_0 A_n + \ldots + a_r A_{n+r} = 0 \tag{5}$$

for all $n \geqslant N$. *Then there are numbers* $\lambda_1, \ldots, \lambda_r$ *such that for all* $n \geqslant N$

$$A_n = \lambda_1 \alpha_1^n + \ldots + \lambda_r \alpha_r^n. \tag{6}$$

Proof. By Lemma 1 there are certainly $\lambda_1, \ldots, \lambda_r$ such that (6) is true for $N \leqslant n < N + r$. But the right-hand side of (6) clearly satisfies (5), and (5) determines $A_{N+r}, A_{N+r+1}, \ldots$ uniquely once $A_N, \ldots, A_{N+r-1}$ are known.

LEMMA 3. *Suppose that there is some $\mu \neq 0$ in the field of α such that μ, $\mu\alpha$, $\mu\alpha^2$, ... are all equal to polynomials in α of degree $r-1$ with rational integer coefficients. Then α is an algebraic integer.*

Proof. Using ideals it is easy to deduce a contradiction from the assumption that some prime ideal occurs in α to a negative power. The following proof does not use ideals.

Let $\mathfrak{B}$ be the set of all numbers β which are the sum of a finite number of expressions $c\mu\alpha^n$ ($n>0$, c rational integers). By hypothesis $\beta = b_0 + b_1\alpha + ... + b_{r-1}\alpha^{r-1}$, where $b_0, ..., b_{r-1}$ are integers. Clearly the set of all $(b_0, ..., b_{r-1})$ with $\beta \in \mathfrak{B}$ is a module in the sense of Appendix A. Let $\beta^{(1)}, ..., \beta^{(s)}$ ($s \leq r$) correspond in the obvious way to a basis of the module, so that $\beta^{(j)} \in \mathfrak{B}$ ($1 \leq j \leq s$) and every $\beta \in \mathfrak{B}$ has the shape

$$\beta = c_1\beta^{(1)} + ... + c_s\beta^{(s)} \quad (c_j, \text{ rational integers}).$$

Clearly $\alpha\beta \in \mathfrak{B}$ if $\beta \in \mathfrak{B}$; so in particular

$$\alpha\beta^{(j)} = \sum_t c_{jt}\beta^{(t)} \quad (t, j = 1, ..., s),$$

for rational integer c_{jt}. On taking the terms $\alpha\beta^{(j)}$ over to the other side and eliminating $\beta^{(1)}, ..., \beta^{(s)}$ we have $\det(\alpha\delta_{jt} - c_{jt}) = 0$ where $\delta_{jt} = 1$ if $t = j$ but $= 0$ otherwise. This is an equation for α of degree $s \leq r$ with rational integer coefficients, the highest coefficient being 1. Thus $s = r$ and α is an algebraic integer.

Proof of Theorem I. Write

$$\lambda\alpha^n = A_n + \epsilon_n, \tag{7}$$

where A_n is an integer and

$$|\epsilon_n| = \|\lambda\alpha^n\| \leq \tfrac{1}{2}, \quad \epsilon_n \to 0 \quad (n \to \infty). \tag{8}$$

From (1), (7), (8) we have

$$a_0 A_n + a_1 A_{n+1} + ... + a_r A_{n+r}$$
$$= \lambda\alpha^n(a_0 + a_1\alpha + ... + a_r\alpha^r) - a_0\epsilon_n - a_1\epsilon_{n+1} - ... - a_r\epsilon_{n+r}$$
$$= -a_0\epsilon_n - ... - a_r\epsilon_{n+r} \to 0. \tag{9}$$

Since the left-hand side of (9) is an integer it is 0 for all $n \geq$ some N.

By Lemma 2, (6) holds for all $n \geqslant N$, where λ_1 is not necessarily λ. Let $m \geqslant N$. On solving for $\lambda_1, \ldots, \lambda_n$ from (6) with

$$n = m, m+1, \ldots, m+r-1,$$

we have, by Lemma 1,

$$\delta_j \alpha_j^m \lambda_j = \sum_{0 \leqslant k < r} \beta_{jk} A_{m+k} \quad (1 \leqslant j \leqslant r). \tag{10}$$

Hence λ_j is in the field of α_j since δ_j, β_{jk} are; and indeed the λ_j are conjugates since the δ_j, β_{jk} are. If any λ_j vanished they would thus all vanish and $A_n = 0$ (all $n \geqslant N$) in contradiction to (7). Hence

$$\lambda_j \neq 0 \quad (1 \leqslant j \leqslant r). \tag{11}$$

The right-hand side of (10) with $j = 1$ is a polynomial in $\alpha = \alpha_1$ of degree $r-1$ with rational integer coefficients, by (3'), and $\delta_1 \lambda_1 \alpha_1^N = \mu$ (say) $\neq 0$. Hence Lemma 3 applies and α is an algebraic integer.

By (6) and (7) we have

$$\epsilon_n = (\lambda - \lambda_1) \alpha_1^n - \lambda_2 \alpha_2^n - \ldots - \lambda_r \alpha_r^n \quad (\text{all } n \geqslant N). \tag{12}$$

Hence, by Lemma 1 again,

$$\delta_1 (\lambda - \lambda_1) \alpha_1^m = \sum_{0 \leqslant k < r} \beta_{1k} \epsilon_{m+k} \quad (\text{all } m \geqslant N)$$

$$\to 0 \quad (m \to \infty)$$

by (8). Thus $\lambda = \lambda_1$ since $\alpha_1 > 1$. Similarly $\delta_j \lambda_j \alpha_j^m \to 0$ $(j > 1)$ by (12), (8). Thus $|\alpha_j| < 1$ $(j > 1)$ by (11).

3. Proof of Theorem II.

An infinite sequence of numbers z_n $(0 \leqslant n < \infty)$ is said to have a RECURRENCE RELATION if there are constants $c_0, \ldots, c_{r-1}$ such that

$$z_{n+r} = c_{r-1} z_{n+r-1} + \ldots + c_0 z_n \tag{1}$$

for all sufficiently large n. We first show that if all the z_n are rational then $c_0, \ldots, c_{r-1}$ can be taken without loss of generality to be rational. For by Lemma 2, Corollary, of Chapter III there are numbers $\mu_1 = 1, \mu_2, \ldots, \mu_l$ for some $l \leqslant r$ which are linearly independent over the rationals and on which $1, c_0, \ldots, c_{r-1}$ are linearly dependent; say

$$c_j = c_j^* + \sum_{1 < k \leqslant l} d_{kj} \mu_k, \tag{2}$$

with rational c_j^*, d_{kj}. If $z_n, \ldots, z_{n+r}$ are all rational we may express the c_j by (2) and then equate the coefficients of $\mu_1 = 1$ on both sides, i.e.

$$z_{n+r} = c_{r-1}^* z_{n+r-1} + \ldots + c_0^* z_n.$$

This is of the type required.

THEOREM IV. *Let z_n $(0 \leqslant n < \infty)$ be a sequence of real numbers and A_n $(0 \leqslant n < \infty)$ a sequence of rational integers with*

$$\Sigma \, | \, z_n - A_n \, |^2 < \infty. \tag{3}$$

Then if the z_n have a recurrence relation so have the A_n (but not necessarily the same relation).

Deduction of Theorem II. We first show how Theorem II follows from Theorem IV. Let λ, α be as in Theorem II and define the integer A_n by $| \, A_n - \lambda \alpha^n \, | = \| \, \lambda \alpha^n \, \|$. The sequence $z_n = \lambda \alpha^n$ has the recurrence relation $z_{n+1} = \alpha z_n$. By Theorem IV and the hypothesis (1·4) of Theorem II the A_n satisfy

$$A_{n+r} = c_{r-1} A_{n+r-1} + \ldots + c_0 A_n \tag{4}$$

for large enough n, where $c_{r-1}, \ldots, c_0$ may be assumed to be rational. On putting $\lambda \alpha^m = A_m + \epsilon_m$ $(n \leqslant m \leqslant n+r)$ in (4) and dividing by $\lambda \alpha^n \neq 0$ we have

$$\alpha^r - c_{r-1} \alpha^{r-1} - \ldots - c_0$$
$$= (\lambda \alpha^n)^{-1} (\epsilon_{n+r} - c_{r-1} \epsilon_{n+r-1} - \ldots - c_0 \epsilon_n)$$
$$\to 0 \quad (n \to \infty).$$

i.e. $\alpha^r - c_{r-1} \alpha^{r-1} - \ldots - c_0 = 0$. Thus α is algebraic and Theorem I applies.

Theorem IV depends on two general lemmas.

LEMMA 4. *A necessary and sufficient condition that the sequence $z_0, z_1, \ldots$ of real numbers have a recurrence relation is that*

$$D_n = \det (z_{i+j}) \quad (0 \leqslant i, j \leqslant n)$$

vanish for all sufficiently large n.

Proof. If a recurrence relation exists there is a linear relation between the rows of D_n for large enough n and then $D_n = 0$. It remains to show that a recurrence relation exists if D_n vanishes for all sufficiently large n. If D_n vanishes for all n then

$$0 = z_0 = z_1 = \ldots = z_n = \ldots$$

in order. Otherwise there is an $r \geqslant 1$ such that

$$D_{r-1} \neq 0, \quad D_n = 0 \quad \text{(all } n \geqslant r). \tag{5}$$

Since $D_r = 0$ there is a linear relation between the rows of the corresponding matrix, say

$$c_r z_{n+r} + c_{r-1} z_{n+r-1} + \ldots + c_0 z_n = 0 \quad (0 \leqslant n \leqslant r),$$

where $c_r, \ldots, c_0$ do not all vanish. If $c_r = 0$ we should have a relation between the rows of D_{r-1}, i.e. $D_{r-1} = 0$ contrary to hypothesis. We may thus suppose that $c_r = -1$. Then (1) holds for $0 \leqslant n \leqslant r$ and we shall show that it holds for all n. Put

$$R_n = z_n - c_{r-1} z_{n-1} - \ldots - c_0 z_{n-r} \quad (n \geqslant r).$$

Then $R_n = 0 \ (r \leqslant n \leqslant 2r)$. Suppose that we know that

$$R_n = 0 \quad (r \leqslant n < N), \tag{6}$$

where $N > 2r$. On replacing the jth column $(j > r)$ of D_{N-r} by $(j$th column$) - c_{r-1}(\overline{j-1}$th column$) \ldots - c_0(\overline{j-r}$th column$)$ we replace the z_{i+j} in position (i, j) of D_{N-r} by R_{i+j} for $j \geqslant r$. Hence $D_{N-r} = \pm D_{r-1} R_N^{N-2r+1}$ on evaluating the determinant using (6). Finally, $R_N = 0$ by (5), and, by induction, $R_n = 0$ for all $n \geqslant r$.

LEMMA 5. *Let* $\Sigma \alpha_{ij} x_i x_j \geqslant 0 \ (\alpha_{ij} = \alpha_{ji})$ *for all* $(x_1, \ldots, x_n)$. *Then*

$$0 \leqslant \det (\alpha_{ij}) \leqslant \Pi \alpha_{ii}.$$

Proof. On successively 'completing the square' and changing the order of $(x_1, \ldots, x_n)$ if need be, we have

$$\Sigma \alpha_{ij} x_i x_j = \beta_1 (x_1 + \lambda_{12} x_2 + \ldots + \lambda_{1n} x_n)^2$$
$$+ \beta_2 (x_2 + \lambda_{23} x_3 + \ldots + \lambda_{2n} x_n)^2$$
$$+ \ldots + \beta_n x_n^2,$$

where the λ_{ij} are real and $\beta_i \geqslant 0 \ (1 \leqslant i \leqslant n)$. Clearly

$$\det (\alpha_{ij}) = \Pi \beta_i \quad \text{and} \quad \alpha_{ii} = \beta_i + \sum_{k<i} \beta_k \lambda_{ki}^2 \geqslant \beta_i.$$

COROLLARY. (Hadamard.) *Let β_{ij} be any n^2 real numbers and*

$$M_i = (\sum_j \beta_{ij}^2)^{\frac{1}{2}};$$

then $$|\det(\beta_{ij})| \leqslant M_1 M_2 \ldots M_n.$$

Note. This is the n-dimensional generalization of the fact that the volume of a parallelepiped is at most the product of the edges.

Proof. $\sum \alpha_{ij} x_i x_j = \sum_j (\sum_i \beta_{ij} x_i)^2 \geqslant 0$ for all $(x_1, \ldots, x_n)$ and $\det(\alpha_{ij}) = (\det(\beta_{ij}))^2$; $\alpha_{ii} = \sum_j \beta_{ij}^2 = M_i^2$.

Proof of Theorem IV. Let

$$\Delta_n = \det(A_{i+j}) \quad (0 \leqslant i, j \leqslant n), \tag{7}$$

and let z_n satisfy (1), say, for $n \geqslant N - r$. For $n \geqslant N$ write

$$\epsilon_n = A_n - c_{r-1} A_{n-1} - \ldots - c_0 A_{n-r}$$
$$= (A_n - z_n) - c_{r-1}(A_{n-1} - z_{n-1}) - \ldots - c_0(A_{n-r} - z_{n-r}).$$

Then by Schwarz's inequality†

$$\epsilon_n^2 \leqslant d \sum_{n-r \leqslant m \leqslant n} (A_m - z_m)^2, \quad d = 1 + c_{r-1}^2 + \ldots + c_0^2,$$

and so $\sum \epsilon_n^2 < \infty$ by (3). For $n \geqslant 2N$ write

$$\eta_n = \epsilon_n - c_{r-1} \epsilon_{n-1} - \ldots - c_0 \epsilon_{n-r},$$

so that similarly $\sum \eta_n^2 < \infty$. By operating on the rows of (7) we may replace A_{i+j} in position (i, j) by ϵ_{i+j} for all $j \geqslant N$ and all i. By operating similarly on the columns we have

$$\Delta_n = \det(\delta_{ij}), \delta_{ij} = \begin{cases} A_{i+j} & \text{if } i < N, j < N, \\ \eta_{i+j} & \text{if } i \geqslant N, j \geqslant N, \\ \epsilon_{i+j} & \text{otherwise.} \end{cases}$$

For $i < N$

$$\sum_j \delta_{ij}^2 \leqslant \sum_{0 \leqslant j \leqslant 2N-2} A_j^2 + \sum_{j \geqslant N} \epsilon_j^2 = \mu_i^2 \quad \text{(say)},$$

and, for $i \geqslant N$

$$\sum_j \delta_{ij}^2 \leqslant \sum_{j \geqslant i} (\epsilon_j^2 + \eta_j^2) = \mu_i^2 \quad \text{(say)}.$$

† See footnote, p. 72.

Then μ_i is independent of n and $\mu_i \to 0$ $(i \to \infty)$ since $\Sigma \eta_j^2$, $\Sigma \epsilon_j^2$ converge. Hence, by Lemma 5, Corollary,

$$|\Delta_n| \leqslant \mu_1 \dots \mu_n \to 0 \quad (n \to \infty).$$

But Δ_n is an integer by (7), so $\Delta_n = 0$ for all sufficiently large n. Thus Lemma 4 applies and A_n has a recurrence relation.

4. Proof of Theorem III. This depends on showing that to each PV-number α there is a $\lambda > 0$ such that λ and $\Sigma \parallel \lambda \alpha^n \parallel^2$ are both not too big. Throughout this section we put $i = \sqrt{(-1)}$ and use a bar (⁻) to denote the complex conjugate; so $|z|^2 = z\bar{z}$.

LEMMA 6. *Suppose that* $\sum_{n \geqslant 0} \beta_n$ *is an absolutely convergent series of real or complex numbers. Then*

$$\int_0^{2\pi} |\sum_{n \geqslant 0} \beta_n e^{in\theta}|^2 d\theta = 2\pi \sum_{n \geqslant 0} |\beta_n|^2.$$

Proof. Clear on expanding and integrating term by term. All the interchanges of limiting processes are justified by the absolute convergence.

LEMMA 7. *Let* z, β *be real or complex numbers with* $|z| \leqslant 1$. *Then*

$$|z - \bar{\beta}| \leqslant |1 - \beta z| \quad \textit{if} \quad |\beta| < 1,$$
$$|z - \bar{\beta}| \geqslant |1 - \beta z| \quad \textit{if} \quad |\beta| > 1.$$

In both cases there is equality if and only if $|z| = 1$.

Proof. This follows at once from the simple identity

$$|1 - \beta z|^2 - |z - \bar{\beta}|^2 = (1 - |\beta|^2)(1 - |z|^2).$$

LEMMA 8. (Fatou.) *Suppose that* $\phi(m, n) \geqslant 0$ *for* $n \geqslant 0$, $m \geqslant 0$ *and that* $\phi^*(n) = \lim_{m \to \infty} \phi(m, n)$ *exists* $(n \geqslant 0)$. *Then*[†]

$$\sum_{n \geqslant 0} \phi^*(n) \leqslant \liminf_{m \to \infty} \sum_{n \geqslant 0} \phi(m, n).$$

Proof.

$$\sum_{n \leqslant N} \phi^*(n) = \lim_{m \to \infty} \sum_{n \leqslant N} \phi(m, n) \leqslant \liminf_{m \to \infty} \sum_{n < \infty} \phi(m, n)$$

† For the notation 'lim inf' see p. x.

for any $N \geqslant 0$, and

$$\sum_{n \geqslant 0} \phi^*(n) = \lim_{N \to \infty} \sum_{n \leqslant N} \phi^*(n).$$

LEMMA 9. *Let α be a PV-number. Then there is a real $\lambda \neq 0$ such that*

$$|\lambda| < \alpha, \quad \sum_n \| \lambda \alpha^n \|^2 < 4\alpha^2/(\alpha - 1)^2. \tag{1}$$

Proof. Let α of degree r with conjugates $\alpha = \alpha_1, \ldots, \alpha_r$ satisfy the irreducible equation

$$f(\alpha) = 0, \quad f(z) = z^r + a_{r-1} z^{r-1} + \ldots + a_0, \tag{2}$$

where $a_{r-1}, \ldots, a_0$ are integers.

We dispose first of an anomalous case:

$$r = 2, \quad a_0 = \pm 1, \quad \text{so} \quad \alpha_2 = \pm \alpha^{-1}. \tag{3}$$

For (3) we put $\quad\quad\quad\quad \lambda = 1 < \alpha.$

Then $\quad\quad\quad\quad \| \lambda \alpha^n \| \leqslant | \alpha^n - T(\alpha^n) | = | \alpha_2^n | = \alpha^{-n},$

and so

$$\sum \| \lambda \alpha^n \|^2 \leqslant \sum_{n \geqslant 0} \alpha^{-2n} = \alpha^2/(\alpha^2 - 1) < 4\alpha^2/(\alpha - 1)^2.$$

We can now exclude (3). Write

$$g(z) = a_0 z^r + a_1 z^{r-1} + \ldots + 1 = z^n f(z^{-1}), \tag{4}$$

and put

$$h(z) = \frac{f(z)}{g(z)} = \sum_{n \geqslant 0} A_n z^n, \tag{5}$$

the expansion being convergent for sufficiently small z. The A_n are rational integers by (2) and (4). Since $f(z)$ is irreducible, either $f(z)$ and $g(z)$ have no root in common or $g(z) = a_0 f(z)$. If $r > 2$ the second alternative is impossible since only one of the roots α_j of $f(z)$ but $r - 1$ of the roots α_j^{-1} of $g(z)$ lie outside $|z| < 1$; and if $r = 2$ the second alternative leads to the excluded case (3). We may thus assume that

$$\alpha_j \alpha_l \neq 1 \quad (1 \leqslant j, l \leqslant r). \tag{6}$$

On factorizing $f(z), g(z)$ we have

$$h(z) = \prod_{1 \leqslant j \leqslant r} \left(\frac{z - \bar{\alpha}_j}{1 - \alpha_j z} \right), \tag{7}$$

since complex roots occur in conjugate pairs. There is a partial fraction expansion

$$h(z) = \frac{1}{a_0} + \sum_{1 \leqslant j \leqslant r} \frac{\lambda_j}{1 - \alpha_j z},$$

where

$$\lambda_j = \lim_{z \to \alpha_j^{-1}} (1 - \alpha_j z) h(z)$$

$$= (\alpha_j^{-1} - \bar{\alpha}_j) \prod_{l \neq j} \left(\frac{\alpha_j^{-1} - \bar{\alpha}_l}{1 - \alpha_j^{-1} \alpha_l} \right)$$

$$\neq 0$$

by (6). In particular, $\lambda = \lambda_1$ is real and

$$0 < |\lambda| < |\alpha^{-1} - \alpha| < \alpha, \tag{8}$$

since

$$|\alpha^{-1} - \bar{\alpha}_l| < |1 - \alpha^{-1} \alpha_l| \quad (l \neq 1),$$

by Lemma 7 with $z = \alpha^{-1}$, $\beta = \alpha_l$. The power series

$$F(z) = \sum_{n \geqslant 0} (A_n - \lambda \alpha^n) z^n = h(z) - \frac{\lambda}{1 - \alpha z}$$

$$= \frac{1}{a_0} + \sum_{j > 1} \frac{\lambda_j}{1 - \alpha_j z}$$

is absolutely convergent for $z = 1$ since $|\alpha_j| < 1$ $(j \neq 1)$. But

$$|h(z)| = 1 \quad \text{if} \quad |z| = 1,$$

by (7) and Lemma 7; and

$$\left| \frac{\lambda}{1 - \alpha z} \right| < \frac{|\lambda|}{\alpha - 1} < \frac{\alpha}{\alpha - 1} \quad \text{if} \quad |z| = 1,$$

by (8). Hence

$$|F(z)| < 1 + \frac{\alpha}{\alpha - 1} < \frac{2\alpha}{\alpha - 1} \quad \text{if} \quad |z| = 1.$$

Finally, by Lemma 6

$$\sum_{n \geqslant 0} \| \lambda \alpha^n \|^2 \leqslant \sum_{n \geqslant 0} |A_n - \lambda \alpha^n|^2$$

$$= \frac{1}{2\pi} \int_0^{2\pi} |F(e^{i\theta})|^2 \, d\theta$$

$$< \left(\frac{2\alpha}{\alpha - 1} \right)^2.$$

COROLLARY. *We may suppose, further, that* $1 \leqslant \lambda < \alpha$.

Proof. We may suppose that $\lambda > 0$ since $-\lambda$ satisfies (1) if λ does. By (1) there is an integer $N \geqslant 0$ such that

$$\lambda' = \lambda \alpha^N < \alpha \leqslant \lambda \alpha^{N+1}.$$

Then $1 \leqslant \lambda' < \alpha$ and

$$\sum_{n \geqslant 0} \| \lambda' \alpha^n \|^2 = \sum_{n \geqslant N} \| \lambda \alpha^n \|^2 \leqslant \sum_{n \geqslant 0}.$$

Proof of Theorem III. Let $\alpha(m)$ $(m = 1, 2, \ldots)$ be a sequence of PV-numbers and $\beta = \lim \alpha(m)$. We must show that β is a PV-number. The corresponding numbers $\lambda(m)$ given by Lemma 9, Corollary, are bounded, and so, by taking a subsequence instead of the original sequence if need be, we may suppose also that $\mu = \lim \lambda(m)$ exists. Clearly $1 \leqslant \mu \leqslant \beta$, in particular $\mu \neq 0$. If $\beta > 1$ we have, by Lemmas 8, 9

$$\sum_{n \geqslant 0} \| \mu \beta^n \|^2 \leqslant \liminf_{m \to \infty} \sum_{n \geqslant 0} \| \lambda(m) \alpha^n(m) \|^2$$

$$\leqslant \liminf_{m \to \infty} \frac{4\alpha^2(m)}{(\alpha(m) - 1)^2}$$

$$= \frac{4\beta^2}{(\beta - 1)^2} < \infty;$$

and β is a PV-number by Theorem II.

Since 1 is not a PV-number it remains only to show that $\beta \neq 1$. If α is a PV-number then, from the definition, so is α^k for any integer $k > 0$. Let γ be any number > 1 not a PV-number (say $\gamma = \frac{3}{2}$). Then $\alpha(m) < \gamma$ for all sufficiently large m if $\beta = 1$; and for such m we may choose an integer $k(m) > 0$ such that

$$\delta(m) = (\alpha(m))^{k(m)} < \gamma \leqslant \delta(m) \alpha(m).$$

The $\delta(m)$ are PV-numbers and $\gamma = \lim \delta(m)$. This contradicts what has already been proved, so $\beta = 1$ cannot occur.

NOTES

Since the set of PV-numbers is closed it must have a least member. In fact the least member and the least limit-point are known. For latest information on this and other topics see DUFRESNOY & PISOT (1954) and for generalizations see PISOT (1946), GEL'FOND (1941), KELLY (1950) and SAMET (1953).

DEVELOPMENTS SINCE 1956

These notes are intended merely to draw attention to some of the papers and review articles that have appeared since the book was originally written and to repair one or two omissions. They are not intended to be exhaustive. References to the supplementary bibliography have an asterisk attached to the date.

I have treated some of the topics whose omission is noted in the Preface in a later book (CASSELS (1959*)). The book of NIVEN (1963*) has a different choice of material from this tract and is more elementary. The Russian books KHINTCHINE (1939) and GEL'FOND (1952) have now been done into American: KHINCHIN (1964*) and GELFOND (1960*).

CHAPTER I. §4. For more on asymmetric generalizations see EGGAN (1961*) and EGGAN & NIVEN (1961*).

§5. For a much stronger result in some cases see DAVENPORT (1962*).

CHAPTER II. §2. Some of the work of CASSELS & SWINNER-TON-DYER (1955) mentioned in the notes has been extended by PECK (1961*).

CHAPTER III. §2. There is an analogue of Theorem I for imaginary quadratic fields. The correct constant to replace $\frac{1}{4}$ on the right-hand side of (1) was conjectured by PERRON (1946*) and proved by BIRCH (1957*) except for a finite number of fields, several special cases having been done earlier. The result of CASSELS (1954a) mentioned in the notes has been carried much further by DESCOMBES (1956*).

CHAPTER IV. CIGLER & HELMBERG (1961*) give an encyclopaedic report on recent work on the many generalizations of uniform distribution. See also DAVENPORT, ERDÖS & LEVEQUE (1963*) and DAVENPORT & LEVEQUE (1963*).

CHAPTER V. §6. It is now known that the constant Γ_{11} of Theorem XI satisfies $12 \cdot 92 < \Gamma_{11}^{-1} < 30 \cdot 7$, see ENNOLA (1958*) and PITMAN (1960*).

CHAPTER VI. For further generalizations and applications of Roth's theorem see LEVEQUE (1956*), MAHLER (1961* and 1963*) and LANG (1962*) and for 'effective' results see BAKER (1964*).

CHAPTER VII. ERDÖS (1959*), LEVEQUE (1958*) and SCHMIDT (1960*) have independently given quantitative forms of Theorem I. For other generalizations see GALLAGHER (1961* and 1962*) and SZÜSS (1958*). For the connection of metrical theory with ergodic theory see RYLL-NARDZEWSKI (1951*), CIGLER & HELMBERG (1961*) and POSTNIKOV (1960*) and also the popular account in KAC (1959*). For sets of fractional dimension see EGGLESTON (1951*) and ROGERS (1964*).

CHAPTER VIII. For a fuller discussion see SALEM (1963*) and the complete bibliography in PISOT (1962*).

APPENDIX B. See also CASSELS (1959*).

BASES IN CERTAIN MODULES

We shall call a set $\mathfrak{M}$ of vectors of some dimension n a MODULE if $\mathbf{x}^{(1)} \pm \mathbf{x}^{(2)}$ belong to $\mathfrak{M}$ whenever $\mathbf{x}^{(1)}$ and $\mathbf{x}^{(2)}$ do so. In particular, $(\mathbf{x}^{(1)} = \mathbf{x}^{(2)})$, the vector $\mathbf{0}$ belongs to $\mathfrak{M}$. By induction

$$y = a_1 \mathbf{x}^{(1)} + \ldots + a_m \mathbf{x}^{(m)} \tag{1}$$

is in $\mathfrak{M}$ whenever $\mathbf{x}^{(1)}, \ldots, \mathbf{x}^{(m)}$ are and $a_1, \ldots, a_m$ are rational integers. We shall say that $\mathbf{x}^{(1)}, \ldots, \mathbf{x}^{(m)}$ is a BASIS for the module if (i) every vector of the module has the shape (1) and (ii) the only solution of (1) with $\mathbf{y} = \mathbf{0}$ in integers $a_1, \ldots, a_m$ is

$$a_1 = \ldots = a_m = 0.$$

Then, clearly, the representation (1) is always unique.

We shall be concerned only with modules all the vectors of which have integral co-ordinates.

LEMMA 1. *If all the vectors of a module $\mathfrak{M}$ have rational integer co-ordinates then $\mathfrak{M}$ has a basis, provided that it contains a vector other than $\mathbf{0}$.*

Proof. We shall use induction. We suppose the lemma is true for $(n-1)$-dimensional vectors and prove it for n-dimensional vectors. By suitably permuting the order of the co-ordinates if necessary, we may suppose that there is a vector

$$\mathbf{x}^{(1)} = (x_{11}, \ldots, x_{1n}), \quad x_{11} \neq 0$$

in the module. We choose $\mathbf{x}^{(1)}$ so that the integer $|x_{11}| \neq 0$ is as small as possible. If $\mathbf{y} = (y_1, \ldots, y_n)$ is any other vector in $\mathfrak{M}$ there is an integer a_1 such that $|y_1 - a_1 x_{11}| < |x_{11}|$. Then

$$\mathbf{y}' = \mathbf{y} - a_1 \mathbf{x}^{(1)} \in \mathfrak{M}.$$

The absolute value of first co-ordinate of $\mathbf{y}'$ is less than $|x_{11}|$; so by the definition of $\mathbf{x}^{(1)}$ we have $\mathbf{y}' = (0, y_2', \ldots, y_n')$ for certain integers $y_2', \ldots, y_n'$. The vectors $\mathbf{y}'$ of this shape clearly form an $(n-1)$-dimensional module† $\mathfrak{M}'$ and so, by the hypothesis of the

† More precisely, the vectors $(y_2', \ldots, y_n')$ do so.

induction, it has a basis, say, $\mathbf{x}^{(2)}, \ldots, \mathbf{x}^{(m)}$ for some $m \geqslant 1$ (where $m = 1$ means that $\mathfrak{M}'$ consists of $\mathbf{0}$ only). Then

$$\mathbf{y} = a_1 \mathbf{x}^{(1)} + \mathbf{y}' = a_1 \mathbf{x}^{(1)} + \ldots + a_m \mathbf{x}^{(m)}$$

for integers $a_1, \ldots, a_m$. On the other hand, $\mathbf{0} = a_1 \mathbf{x}^{(1)} + \ldots + a_m \mathbf{x}^{(m)}$ implies $a_1 = 0$ on comparing the first co-ordinates; then $a_2 = \ldots = a_m = 0$ since $\mathbf{x}^{(2)}, \ldots, \mathbf{x}^{(m)}$ are a basis for $\mathfrak{M}'$. Hence $\mathbf{x}^{(1)}, \ldots, \mathbf{x}^{(m)}$ is a basis.

This proves the lemma, since the proof for $n = 1$ is a trivially simpler version of the deduction from $n - 1$ to n.

COROLLARY. *After a suitable permutation of the co-ordinates the basis may be put in the form*

$$\mathbf{x}^{(j)} = (0, \ldots, 0, x_{jj}, \ldots, x_{jn}), \quad x_{jj} \neq 0 \quad (1 \leqslant j \leqslant m),$$

(that is, $x_{jk} = 0$ if $k < j$). If $m = n$ no permutation is necessary.

Proof. Clear.

The special module $\mathfrak{M}_0$ of all integer vectors has one basis $\mathbf{e}^{(j)} = (0, \ldots, 0, 1, 0, \ldots, 0)$, where the 1 is in the jth place $(1 \leqslant j \leqslant n)$, but there are others:

LEMMA 2. *The necessary and sufficient condition that*

$$\mathbf{x}^{(j)} = (x_{j1}, \ldots, x_{jn}) \quad (1 \leqslant j \leqslant n) \tag{2}$$

with integral x_{jk} be a basis for the module $\mathfrak{M}_0$ of all integer vectors is that

$$\det (x_{jk}) = \pm 1. \tag{3}$$

Proof. If the $\mathbf{x}^{(j)}$ are a basis, there are rational integers d_{kj} such that $\mathbf{e}^{(j)} = \sum_k d_{jk} \mathbf{x}^{(k)}$, that is

$$\sum_k d_{jk} x_{kl} = 1 \quad \text{if} \quad j = l,$$
$$= 0 \quad \text{if} \quad j \neq l.$$

Hence $\det (d_{jk}) \det (x_{kl}) = 1$. Since the determinants are integers, this implies (3). Conversely, if (3) holds and $\mathbf{y}$ has integer co-ordinates the solution of $\mathbf{y} = \sum_j a_j \mathbf{x}^{(j)}$ by the usual determinantal method gives integers a_j. Further, $\mathbf{0} = \Sigma a_j \mathbf{x}^{(j)}$ implies

$$a_1 = \ldots = a_n = 0,$$

since these are equations with non-vanishing determinant.

LEMMA 3. *Let* $\mathbf{y}^{(1)}, \ldots, \mathbf{y}^{(n)}$ *be vectors of the module* $\mathfrak{M}_0$ *of all n-dimensional integral vectors and suppose that* $\Sigma b_j \mathbf{y}^{(j)} = 0$ *implies* $b_j = 0$ $(1 \leqslant j \leqslant n)$ *(i.e. the vectors are linearly independent). Then there is a basis* $\mathbf{x}^{(j)}$ *for* $\mathfrak{M}_0$ *such that*

$$\mathbf{y}^{(j)} = c_{j1} \mathbf{x}^{(1)} + \ldots + c_{jj} \mathbf{x}^{(j)} \quad (1 \leqslant j \leqslant n), \tag{4}$$

for integers c_{jk} *with* $c_{jk} = 0$ *if* $k > j$. *Further,* $c_{jj} \neq 0$.

Proof. Let $d = \det(y_{jk})$, so $d \neq 0$ is an integer. Then every integer vector $\mathbf{x}$ may be put in the form

$$d\mathbf{x} = \Sigma t_j \mathbf{y}^{(j)}$$

for integers $t_1, \ldots, t_n$. The set of integral vectors $\mathbf{t} = (t_1, \ldots, t_n)$ which may arise in this way clearly form a module $\mathfrak{M}$ and so, by Lemma 1, Corollary (with the order of the suffixes reversed), there is a basis
$$\mathbf{t}^{(j)} = (t_{j1}, \ldots, t_{jj}, 0, \ldots, 0) \quad (t_{jj} \neq 0),$$

for $\mathfrak{M}$. Clearly the integral vectors $\mathbf{x}^{(j)}$ defined by

$$d\mathbf{x}^{(j)} = t_{j1} \mathbf{y}^{(1)} + \ldots + t_{jj} \mathbf{y}^{(j)} \tag{5}$$

form a basis for $\mathfrak{M}_0$. On solving (5) successively for $\mathbf{y}^{(1)}, \ldots, \mathbf{y}^{(n)}$ we get equations of the type (4) with rational c_{jk}; in particular $c_{jj} = d/t_{jj} \neq 0$. Finally, the c_{jk} are integers since $\mathbf{x}^{(j)}$ is a basis.

150

TOOLS FROM THE GEOMETRY OF NUMBERS

In the text we have in several places to establish the existence of integers $x_1, \ldots, x_n$ not all 0 satisfying a set of inequalities of one of the types

$$|a_{i1}x_1 + \ldots + a_{in}x_n| \leqslant c_i \text{ or } < c_i \tag{1}$$

for $1 \leqslant i \leqslant$ some m, where the a_{ij} are real numbers and $c_i > 0$. The existence of such integers may be interpreted as the existence of a point with integer co-ordinates other than the origin in the region $\mathscr{R}$ defined by (1), when $x_1, \ldots, x_n$ are treated as ordinary rectangular co-ordinates in n-dimensional Euclidean space.† The simplest form of the theory we shall develop states that there are certainly points with integral co-ordinates other than the origin in $\mathscr{R}$ if $\mathscr{R}$ has a volume $V > 2^n$.

We adopt a vector notation (see p. ix) and denote‡ by $\lambda\mathscr{R}$ the set of points $\lambda\mathbf{x}$ where $\mathbf{x}$ is in $\mathscr{R}$. We are interested only in regions of the very simplest sort indicated above, so we shall not go into deep general questions. In particular we shall assume that all regions considered have a volume (possibly ∞) which has the natural properties.

It is convenient to deal with regions more general than those defined by (1), namely the convex regions symmetric about the origin. A region $\mathscr{R}$ is SYMMETRIC (about $\mathbf{0}$) if $-\mathscr{R} = \mathscr{R}$; i.e. $-\mathbf{x} \in \mathscr{R}$ whenever $\mathbf{x} \in \mathscr{R}$. A region $\mathscr{R}$ is CONVEX if $\lambda\mathbf{x} + \mu\mathbf{y} \in \mathscr{R}$ whenever $\lambda \geqslant 0$, $\mu \geqslant 0$, $\lambda + \mu = 1$ and $\mathbf{x}, \mathbf{y} \in \mathscr{R}$.

The significance of the last definition is that if $\mathscr{R}$ contains $\mathbf{x}$ and $\mathbf{y}$ then it contains the whole of the line segment joining them.

We note that the two definitions are independent of the co-ordinate system and that if $\mathscr{R}$ enjoys either property so does $\lambda\mathscr{R}$ for all λ.

LEMMA 1. *A region $\mathscr{R}$ defined by inequalities of the type* (1) *is convex and symmetric about* $\mathbf{0}$.

† In fact all the properties we discuss are affine invariants.
‡ Thus if $\mathscr{R}$ is defined by (1), then $\lambda\mathscr{R}$ is defined by replacing all the c_i by λc_i.

Proof. Symmetry about 0 is immediate; so we need prove only convexity. Let $\mathbf{x}$, $\mathbf{y}$ be two points of $\mathscr{R}$ and

$$\mathbf{z} = \lambda\mathbf{x} + \mu\mathbf{y}, \quad \lambda \geqslant 0, \quad \mu \geqslant 0, \quad \lambda + \mu = 1.$$

Then

$$|a_{i1}z_1 + \ldots + a_{in}z_n|$$
$$\leqslant \lambda |a_{i1}x_1 + \ldots + a_{in}x_n| + \mu |a_{i1}y_1 + \ldots + a_{in}y_n|$$
$$\leqslant \max(|a_{i1}x_1 + \ldots + a_{in}x_n|, |a_{i1}y_1 + \ldots + a_{in}y_n|).$$

Hence if $\mathbf{x}$, $\mathbf{y}$ satisfy (1) then so does $\mathbf{z}$.

LEMMA 2. *If $\mathscr{R}$ is convex and symmetric about 0 then $\lambda\mathbf{x} \in \mathscr{R}$ whenever $|\lambda| \leqslant 1$ and $\mathbf{x} \in \mathscr{R}$.*

Proof. By symmetry $-\mathbf{x} \in \mathscr{R}$ and so, by convexity,

$$\rho\mathbf{x} + \sigma(-\mathbf{x}) = \lambda\mathbf{x} \in \mathscr{R},$$

where

$$\rho = \tfrac{1}{2}(1 + \lambda) \geqslant 0, \quad \sigma = \tfrac{1}{2}(1 - \lambda) \geqslant 0, \quad \rho + \sigma = 1.$$

LEMMA 3. *If $\mathscr{R}$ is convex and symmetric then $\lambda\mathbf{x} + \mu\mathbf{y} \in \mathscr{R}$ whenever $|\lambda| + |\mu| \leqslant 1$ and $\mathbf{x} \in \mathscr{R}$, $\mathbf{y} \in \mathscr{R}$.*

Note. In geometrical terms this means that if $\mathscr{R}$ contains the points $\mathbf{x}$ and $\mathbf{y}$ then it contains the whole parallelogram with vertices $\pm\mathbf{x}$, $\pm\mathbf{y}$ (and centre 0).

Proof. By Lemma 2,

$$\mathbf{x}' = \eta_1(|\lambda| + |\mu|)\mathbf{x} \in \mathscr{R}, \quad \mathbf{y}' = \eta_2(|\lambda| + |\mu|)\mathbf{y} \in \mathscr{R},$$

where η_1, η_2 are the signs of λ, μ respectively. Hence, by convexity,

$$\lambda\mathbf{x} + \mu\mathbf{y} = \rho\mathbf{x}' + \sigma\mathbf{y}' \in \mathscr{R},$$

where

$$\rho = \frac{|\lambda|}{|\lambda| + |\mu|}, \quad \sigma = \frac{|\mu|}{|\lambda| + |\mu|}, \quad \rho + \sigma = 1.$$

We may now proceed to the main results. The first does not require $\mathscr{R}$ to be convex or symmetric.

THEOREM I. (Blichfeldt.) *Suppose that $\mathscr{R}$ is a region in n-dimensional space of volume $V > 1$ (possibly $V = \infty$). Then there are two distinct points $\mathbf{x}' \in \mathscr{R}$, $\mathbf{x}'' \in \mathscr{R}$ such that $\mathbf{x}'' - \mathbf{x}'$ has integer co-ordinates.*

Proof. For all sets of integers $\mathbf{u} = (u_1, ..., u_n)$ we define $\mathscr{R}_{\mathbf{u}}$ to be the portion of $\mathscr{R}$ in the hypercube

$$u_i \leqslant x_i < u_i + 1 \quad (1 \leqslant i \leqslant n).$$

Denote by $\mathscr{S}_{\mathbf{u}}$ the set of points in the hypercube $0 \leqslant x_i < 1$ obtained from $\mathscr{R}_{\mathbf{u}}$ by the translation $-\mathbf{u}$ (i.e. $\mathscr{S}_{\mathbf{u}}$ is the set of points $\mathbf{x} - \mathbf{u}$ where $\mathbf{x}$ is in $\mathscr{R}_{\mathbf{u}}$). Then $\mathscr{S}_{\mathbf{u}}$ has volume $V_{\mathbf{u}}$ where $\Sigma V_{\mathbf{u}} = V > 1$. Since the hypercube $0 \leqslant x_i < 1$ has volume 1, at least two of the $\mathscr{S}_{\mathbf{u}}$, say $\mathscr{S}_{\mathbf{u}'}$, $\mathscr{S}_{\mathbf{u}''}$, must overlap. Hence there are points $\mathbf{x}'$, $\mathbf{x}''$ of $\mathscr{R}_{\mathbf{u}'}$, $\mathscr{R}_{\mathbf{u}''}$ respectively (and so of $\mathscr{R}$) such that $\mathbf{x}' - \mathbf{u}' = \mathbf{x}'' - \mathbf{u}''$: and then $\mathbf{x}'$, $\mathbf{x}''$ have the property required.

THEOREM II. (Minkowski.) *Let $\mathscr{R}$ be a convex region symmetrical about $\mathbf{0}$ and of volume $V > 2^n$ (possibly $V = \infty$). Then $\mathscr{R}$ contains a point with integer co-ordinates other than $\mathbf{0}$.*

Proof. $\frac{1}{2}\mathscr{R}$ has volume $(\frac{1}{2})^n V > 1$. Hence, by Theorem I there are $\mathbf{x}'$, $\mathbf{x}'' \in \frac{1}{2}\mathscr{R}$ such that $\mathbf{x}' - \mathbf{x}'' = \mathbf{u}$ (say) has integral co-ordinates. But then, by Lemma 3, the point $\frac{1}{2}\mathbf{x}' - \frac{1}{2}\mathbf{x}'' = \frac{1}{2}\mathbf{u} \in \frac{1}{2}\mathscr{R}$, i.e. $\mathbf{u} \in \mathscr{R}$ as required.

The conclusion of Theorem II may cease to hold when $V = 2^n$; as is shown by the $\mathscr{R}$ defined by $|x_i| < 1$ $(1 \leqslant i \leqslant n)$. This has volume 2^n but clearly contains no integral points other than $\mathbf{0}$. However, if $\mathscr{R}$ satisfies certain supplementary conditions then the conclusion continues to hold. We first deal with a special case. A region $\mathscr{R}$ is said to be BOUNDED when all its points lie in a hypercube $|x_i| \leqslant R$ $(1 \leqslant i \leqslant n)$ for a suitable value of R.

LEMMA 4. *The region $\mathscr{R}$ defined by n equations of the type*

$$|a_{i1}x_1 + ... + a_{in}x_n| \leqslant c_i \quad \text{or} \quad < c_i, \tag{2}$$

where $d = |\det(a_{ij})| > 0$, is bounded and has volume

$$V = 2^n d^{-1} c_1 ... c_n.$$

Proof. Write $\xi_i = \Sigma a_{ij} x_j$. Then if α_{ij} is the inverse matrix to a_{ij} and $\mathbf{x} \in \mathscr{R}$ we have

$$\begin{aligned} |x_i| &= |\Sigma \alpha_{ij} \xi_j| \\ &\leqslant \Sigma |\alpha_{ij}| c_j \leqslant \text{some } R \end{aligned}$$

which is independent of $\mathbf{x}$ and i. The value of V follows at once from elementary integration or otherwise.

COROLLARY 1. *A region $\mathcal{R}$ defined by more than n equations of the type (2) is bounded if some set of n out of these equations satisfy the conditions of the lemma.*

COROLLARY 2. *If $\mathcal{R}$ is defined by $m < n$ equations of the type (2) or by n equations of this type with $\det(a_{ij}) = 0$ then $V = \infty$; and $\mathcal{R}$ is not bounded.*

Proofs. Clear.

COROLLARY 3. *Suppose that either $m < n$ or $m = n$ and $\det(a_{ij}) = 0$. Let $c_1, \ldots, c_m$ be any positive numbers, however small. Then there are integers $x_1, \ldots, x_n$ not all 0 such that*

$$|\Sigma a_{ij} x_j| < c_i \quad (1 \leqslant i \leqslant m).$$

Proof. Follows at once from Corollary 2 and Theorem II.

THEOREM III. (Minkowski.) *There are integers x_j not all 0 such that*

$$|\Sigma a_{1j} x_j| \leqslant c_1, \quad |\Sigma a_{ij} x_j| < c_i \quad (2 \leqslant i \leqslant n), \tag{3}$$

provided that

$$c_1 \ldots c_n \geqslant |\det(a_{ij})|. \tag{4}$$

Proof. When $>$ holds in (4) then this follows at once from Theorem II and the last lemma. Suppose now that $=$ holds in (4). By Theorem II for each ϵ in $0 < \epsilon < 1$ we may find integers $\mathbf{x}^{(\epsilon)} \neq \mathbf{0}$ such that

$$|\Sigma a_{1j} x_j^{(\epsilon)}| < c_1 + \epsilon < c_1 + 1, \quad |\Sigma a_{ij} x_j^{(\epsilon)}| < c_i \quad (i \neq 1). \tag{5}$$

By Lemma 4 all satisfy $|x_j^{(\epsilon)}| \leqslant R$ $(1 \leqslant j \leqslant n)$, where R is some number independent of ϵ; and so only a finite number of $\mathbf{x} \neq \mathbf{0}$ can occur as $\mathbf{x}^{(\epsilon)}$. Some integral $\mathbf{x}^{(0)} \neq \mathbf{0}$, say, must occur as $\mathbf{x}^{(\epsilon)}$ with arbitrarily small ϵ. On writing $x_j^{(0)}$ for $x_j^{(\epsilon)}$ in (5) and letting $\epsilon \to 0$ we have what is required.

We now give an extension to $V = 2^n$ of Theorem II. A region $\mathcal{R}$ is CLOSED if whenever all the points $\mathbf{x}^{(m)}$ $(m = 1, 2, \ldots)$ are in $\mathcal{R}$ and $\mathbf{x}^{(0)} = \lim \mathbf{x}^{(m)}$ exists (in the sense that each co-ordinate of $\mathbf{x}^{(m)}$ tends to the corresponding co-ordinate of $\mathbf{x}^{(0)}$) then $\mathbf{x}^{(0)}$ is also in $\mathcal{R}$.

Thus if $\mathcal{R}$ is defined solely by inequalities of the type $|\Sigma a_{ij} x_j| \leqslant c_i$ then $\mathcal{R}$ is closed. Roughly speaking $\mathcal{R}$ is closed if it contains its boundary.

THEOREM IV. (Minkowski.) *Suppose that a region $\mathscr{R}$ is convex, symmetric about the origin, closed and bounded.*† *Then there is a point $\mathbf{x} \neq 0$ with integer co-ordinates in $\mathscr{R}$ provided that its volume $V \geqslant 2^n$.*

Proof. The region $(1+\epsilon)\,\mathscr{R}$ for $0 < \epsilon < 1$ has volume

$$(1+\epsilon)^n V > 2^n.$$

By Theorem II there is an integral point $\mathbf{x}^{(\epsilon)} \neq 0$ in $(1+\epsilon)\mathscr{R}$ and *a fortiori* in $2\mathscr{R}$. By the boundedness of $\mathscr{R}$ only a finite number of integral points may occur as $\mathbf{x}^{(\epsilon)}$ and so one of them, say $\mathbf{x}^{(0)}$, must occur in $(1+\epsilon)\,\mathscr{R}$ for arbitrarily small ϵ. That is,

$$(1+\epsilon)^{-1}\mathbf{x}^{(0)} \in \mathscr{R}$$

for arbitrarily small ϵ. Hence $\mathbf{x}^{(0)} \in \mathscr{R}$ since $\mathscr{R}$ is closed.

We note that Theorem III tells us more than Theorem IV in its special case since the region defined by (3) is not closed.

It‡ is sometimes useful to know more than that a region $\mathscr{R}$ contains one integer point other than the origin. A set of J points $\mathbf{x}^{(1)}, \ldots, \mathbf{x}^{(J)}$ is LINEARLY INDEPENDENT if

$$\mu_1 \mathbf{x}^{(1)} + \ldots + \mu_J \mathbf{x}^{(J)} = 0$$

for numbers $\mu_1, \ldots, \mu_J$ implies $\mu_1 = \ldots = \mu_J = 0$. We shall investigate when $\mathscr{R}$ contains J linearly independent points.

In what follows we shall assume for simplicity that

$$\left.\begin{array}{l} \mathscr{R} \text{ is convex, symmetrical about } \mathbf{0}, \text{ and} \\ \text{closed: it has volume } V, \, 0 < V < \infty. \end{array}\right\} \tag{6}$$

[If $\mathscr{R}$ is not closed one considers instead of $\mathscr{R}$ the set $\overline{\mathscr{R}}$ consisting of $\mathscr{R}$ together with its boundary points.]

For any vector $\mathbf{x}$ we define the DISTANCE FUNCTION $F(\mathbf{x})$ with respect to $\mathscr{R}$ to be the lower bound of the numbers λ such that $\lambda^{-1}\mathbf{x} \in \mathscr{R}$: if there is no such λ we put conventionally§ $F(\mathbf{x}) = \infty$. Then $0 \leqslant F(\mathbf{x}) \leqslant \infty$ and $F(\mathbf{x}) = 0$ only for $\mathbf{x} = 0$ since $\mathscr{R}$ is bounded. For example if $\mathscr{R}$ is defined by $|\Sigma a_{ij} x_j| \leqslant c_i$ then

$$F(\mathbf{x}) = \max_i c_i^{-1} \left| \sum_j a_{ij} x_j \right|.$$

---

† It may be shown that boundedness follows from convexity if $0 < V < \infty$.
‡ The rest of the appendix is required only for Chapter V, §§ 8, 9.
§ We see later that this cannot occur.

It is often more convenient to deal with $F(\mathbf{x})$ instead of with $\mathscr{R}$ directly. Its principal properties are given by the next two lemmas.

LEMMA 5. *The necessary and sufficient condition that* $\mathbf{x} \in \lambda\mathscr{R}$ *for* $\lambda \geqslant 0$ *is that* $\lambda \geqslant F(\mathbf{x})$.

Proof. From the definition $(F(\mathbf{x}))^{-1}\mathbf{x} \in \mathscr{R}$ since $\mathscr{R}$ is closed. By Lemma 2, $\lambda^{-1}\mathbf{x} \in \mathscr{R}$ for $\lambda \geqslant F(\mathbf{x})$. Finally $\lambda^{-1}\mathbf{x} \notin \mathscr{R}$ for $\lambda < F(\mathbf{x})$ by definition.

LEMMA 6. (i) $F(\lambda\mathbf{x}) = |\lambda| F(\mathbf{x})$ *for all vectors* $\mathbf{x}$ *and numbers* λ.
(ii) $F(\mathbf{x}^{(1)} + \mathbf{x}^{(2)}) \leqslant F(\mathbf{x}^{(1)}) + F(\mathbf{x}^{(2)})$ *for any vectors* $\mathbf{x}^{(1)}$, $\mathbf{x}^{(2)}$.

Proof. (i) Trivial.
(ii) Put $\mu_j = F(\mathbf{x}^{(j)})$ so that $\mu_j^{-1}\mathbf{x}^{(j)} \in \mathscr{R}$ $(j = 1, 2)$. Then

$$(\mu_1 + \mu_2)^{-1}(\mathbf{x}^{(1)} + \mathbf{x}^{(2)}) = \frac{\mu_1}{\mu_1 + \mu_2}(\mu_1^{-1}\mathbf{x}^{(1)}) + \frac{\mu_2}{\mu_1 + \mu_2}(\mu_2^{-1}\mathbf{x}^{(2)})$$

is in $\mathscr{R}$ by the definition of convexity; that is

$$F(\mathbf{x}^{(1)} + \mathbf{x}^{(2)}) \leqslant \mu_1 + \mu_2,$$

as required.

Since $V > 0$ there must be n linearly independent points $\mathbf{z}^{(1)}, \ldots, \mathbf{z}^{(n)}$ in $\mathscr{R}$ (not necessarily integral). For any numbers $\mu_1, \ldots, \mu_n$, we have

$$F(\mu_1\mathbf{z}^{(1)} + \ldots + \mu_n\mathbf{z}^{(n)}) \leqslant |\mu_1| F(\mathbf{z}^{(1)}) + \ldots + |\mu_n| F(\mathbf{z}^{(n)})$$
$$\leqslant |\mu_1| + \ldots + |\mu_n|$$

by Lemma 6. Thus $\mathscr{R}$ contains the whole 'generalized octahedron'
$$\mu_1\mathbf{z}^{(1)} + \ldots + \mu_n\mathbf{z}^{(n)}, \quad |\mu_1| + \ldots + |\mu_n| \leqslant 1.$$

In particular, $\lambda\mathscr{R}$ contains any given point if λ is large enough.

By Lemma 5 for each J, $1 \leqslant J \leqslant n$, there is a least λ, say λ_J, such that $\lambda\mathscr{R}$ contains J linearly independent integer points. We call λ_J the Jth SUCCESSIVE MINIMUM of $\mathscr{R}$. Theorem II shows that $\lambda_1^n V \leqslant 2^n$ since for all $\lambda < \lambda_1$ the region $\lambda\mathscr{R}$ has volume $\lambda^n V$ and contains no integral point other than $\mathbf{0}$. By considering the region $\mathscr{R}$ defined by $|x_1| \leqslant M$, $|x_i| \leqslant 1$ $(2 \leqslant i \leqslant n)$ where M is large, which has volume $V = 2^n M$, $\lambda_1 = M^{-1}$, $\lambda_J = 1$ $(2 \leqslant J)$, it is, however, easy to verify that no estimate for λ_J $(2 \leqslant J)$ in terms of V is possible. The following theorem gives an estimate for the product $\lambda_1 \ldots \lambda_n$.

THEOREM V. (Minkowski.) *The successive minima satisfy*

$$2^n/n! \leqslant V\lambda_1 \ldots \lambda_n \leqslant 2^n.$$

Note. Theorem IV follows at once from this since if $V \geqslant 2^n$ we have $\lambda_1^n \leqslant \lambda_1\lambda_2 \ldots \lambda_n \leqslant 1$, $\lambda_1 \leqslant 1$; i.e. $\mathscr{R} = 1\mathscr{R}$ contains an integral point other than $\mathbf{0}$.

Proof. The left-hand inequality is quite straightforward. We choose successively n points $\mathbf{x}^{(1)}, \ldots, \mathbf{x}^{(n)}$ with integer coefficients such that $\mathbf{x}^{(J)}$ lies in $\lambda_J \mathscr{R}$ and is linearly independent of

$$\mathbf{x}^{(1)}, \ldots, \mathbf{x}^{(J-1)}.$$

This is clearly possible. Let $\mathbf{x}^{(J)}$ have co-ordinates $(x_{J1}, \ldots, x_{Jn})$ so that $\det(x_{ji}) \neq 0$ and hence

$$|\det(x_{ji})| \geqslant 1,$$

since the x_{ji} are integers. For any constants $\mu_1, \ldots, \mu_n$ we have

$$F(\mu_1\mathbf{x}^{(1)} + \ldots + \mu_n\mathbf{x}^{(n)}) \leqslant |\mu_1| F(\mathbf{x}^{(1)}) + \ldots + |\mu_n| F(\mathbf{x}^{(n)})$$
$$\leqslant |\mu_1| \lambda_1 + \ldots + |\mu_n| \lambda_n$$

by Lemma 6. Hence

$$\mu_1\mathbf{x}^{(1)} + \ldots + \mu_n\mathbf{x}^{(n)} \tag{7}$$

is in $\mathscr{R}$ provided that

$$|\mu_1| \lambda_1 + \ldots + |\mu_n| \lambda_n \leqslant 1. \tag{8}$$

But it is readily verified that the set of (7) subject to (8) has volume†

$$\frac{2^n |\det(x_{ji})|}{n! \lambda_1 \ldots \lambda_n} \geqslant \frac{2^n}{n! \lambda_1 \ldots \lambda_n}.$$

This volume can be at most the volume V of $\mathscr{R}$ which proves the first half of the theorem.

The proof of the right-hand inequality is much more troublesome. It is convenient to introduce a change of co-ordinates of the type

$$x_i' = t_{i1}x_1 + \ldots + t_{in}x_n, \tag{9}$$

† E.g. by taking $\mu_1, \ldots, \mu_n$ as variables of integration.

where the t_{ij} are integers and

$$\det (t_{ij}) = \pm 1. \tag{10}$$

On solving for the x_i in terms of the x_i' we have

$$x_i = s_{i1}x_1' + \ldots + s_{in}x_n', \tag{11}$$

where the s_{ij} are again integers by (10). Hence (9) takes integer co-ordinates into integer co-ordinates and conversely, so that when we talk of a point with integer co-ordinates it is immaterial whether the old or the new system is meant. As we have already remarked the definitions of convexity and symmetry are independent of the co-ordinate system.

LEMMA 7. *If* $\mathbf{x}^{(1)}, \ldots, \mathbf{x}^{(n)}$ *are* n *linearly independent points with integer co-ordinates there is a co-ordinate change of the type* (9), (10) *such that* $\mathbf{x}^{(i)}$ *has new co-ordinates of the type*

$$(x_{i1}', x_{i2}', \ldots, x_{ii}', 0, \ldots, 0)$$

for $1 \leqslant i \leqslant n$.

Proof. This is only a rewording of Lemmas 2 and 3 of Appendix A. By Lemma 3 (Appendix A) there are n integral vectors $\mathbf{s}^{(i)} = (s_{i1}, \ldots, s_{in})$ forming a basis for the module of all integral vectors and such that $\mathbf{x}^{(i)} = x_{i1}'\mathbf{s}^{(1)} + \ldots + x_{ii}'\mathbf{s}^{(i)}$ for integers x_{ij}'. Since $\det (s_{ij}) = \pm 1$ by Lemma 2 (Appendix A) the transformation (11) with these s_{ij} does what is required.

We shall therefore assume that the $\mathbf{x}^{(i)}$ giving the successive minima have co-ordinates

$$\mathbf{x}^{(i)} = (x_{i1}, \ldots, x_{ii}, 0, \ldots, 0).$$

LEMMA 8. *If* $\mathbf{x}$ *is integral and* $F(\mathbf{x}) < \lambda_J$ *then*

$$x_J = x_{J+1} = \ldots = x_n = 0.$$

Proof. For $\mathbf{x}$ cannot be linearly independent of

$$\mathbf{x}^{(1)}, \ldots, \mathbf{x}^{(J-1)}.$$

COROLLARY. *Let* $\mathbf{x}'' - \mathbf{x}'$ *be integral and*

$$F(\mathbf{x}') < \tfrac{1}{2}\lambda_J, \quad F(\mathbf{x}'') < \tfrac{1}{2}\lambda_J.$$

Then $\qquad\qquad x_j' = x_j'' \quad (J \leqslant j \leqslant n).$

Proof. $\qquad F(\mathbf{x}'' - \mathbf{x}') \leqslant F(\mathbf{x}'') + F(\mathbf{x}') < \lambda_J.$

Proof of Theorem V (*continued*). Put $\mathscr{W}_0(\lambda) = \lambda\mathscr{R}$ and for each integer J $(1 \leqslant J \leqslant n)$ define $\mathscr{W}_J(\lambda)$ to be the set of†

$$(\{x_1\}, ..., \{x_J\}, x_{J+1}, ..., x_n)$$

where $\mathbf{x} \in \lambda\mathscr{R}$. If $\lambda \geqslant \lambda'$ then $\mathscr{W}_J(\lambda)$ contains $\mathscr{W}_J(\lambda')$ since $\lambda\mathscr{R}$ contains $\lambda'\mathscr{R}$: but the difference between the volumes of $\mathscr{W}_J(\lambda)$ and $\mathscr{W}_J(\lambda')$ is clearly at most the difference between the volumes of $\lambda\mathscr{R}$ and $\lambda'\mathscr{R}$, i.e. $(\lambda^n - \lambda'^n) V$. Hence the volume $V_J(\lambda)$ (say) of $\mathscr{W}_J(\lambda)$ increases continuously with λ. Since $\mathscr{W}_n(\lambda)$ lies entirely in the unit cube we have

$$V_n(\lambda) \leqslant 1 \quad \text{(all } \lambda\text{)}. \tag{12}$$

LEMMA 9. $V_n(\lambda) = V_J(\lambda)$ *if* $\lambda \leqslant \tfrac{1}{2}\lambda_{J+1}$ $(J < n)$.

Proof. If $\lambda < \tfrac{1}{2}\lambda_{J+1}$ this follows at once from Lemma 8, Corollary. For $\lambda = \tfrac{1}{2}\lambda_{J+1}$ it is true by continuity.

In particular

$$V_n(\lambda) = V_0(\lambda) = \lambda^n V \quad (\lambda \leqslant \tfrac{1}{2}\lambda_1). \tag{13}$$

LEMMA 10. *Let $\mathscr{S}$ be some region of the unit J-dimensional cube* $0 \leqslant x_j < 1$ $(1 \leqslant j \leqslant J)$ *and let $\mathscr{S}'$ be the set of points* $(\{b_j + x_j\})$ $(1 \leqslant j \leqslant J)$, *where* $b_1, ..., b_J$ *are fixed and* $(x_1, ..., x_J) \in \mathscr{S}$. *Then $\mathscr{S}$ and $\mathscr{S}'$ have the same volume.*

Proof. Clear.

LEMMA 11. $V_J(\lambda) \geqslant (\lambda/\lambda')^{n-J} V_J(\lambda')$ *if* $\lambda \geqslant \lambda'$.

Proof. For any $a_{J+1}, ..., a_n$ denote by $v(a_{J+1}, ..., a_n)$ the J-dimensional volume of the section of $\mathscr{W}_J(\lambda)$ lying in

$$x_{J+1} = a_{J+1}, ..., x_n = a_n,$$

so that $$V_J(\lambda) = \int ... \int v(x_{J+1}, ..., x_n)\, dx_{J+1} ... dx_n. \tag{14}$$

Let $v'(a_{J+1}, ..., a_n)$ be similarly defined with respect to $\mathscr{W}_J(\lambda')$. To prove the lemma, in view of (14), it is clearly enough to prove

$$v\left(\frac{\lambda}{\lambda'} a_{J+1}, ..., \frac{\lambda}{\lambda'} a_n\right) \geqslant v'(a_{J+1}, ..., a_n) \tag{15}$$

for every $(a_{J+1}, ..., a_n)$.

---

† For the notation $\{x\}$ see p. ix.

This is certainly true if the right-hand side is 0. If not, there is some point, say $(a_1, ..., a_J, a_{J+1}, ..., a_n) \in \lambda'\mathscr{R}$ with the correct last $n-J$ co-ordinates. This we keep fixed in the rest of the argument. Now let $(x_1, ..., x_J, a_{J+1}, ..., a_n) \in \mathscr{W}_J(\lambda')$, so that in particular $0 \leqslant x_j < 1$ $(1 \leqslant j \leqslant J)$. Then there are integers $(u_1, ..., u_J)$ such that

$$(x_1 + u_1, ..., x_J + u_J, a_{J+1}, ..., a_n) \in \lambda'\mathscr{R}.$$

Hence

$$\left(\frac{\lambda}{\lambda'} - 1\right)(a_1, ..., a_n) + (x_1 + u_1, ..., x_J + u_J, a_{J+1}, ..., a_n) \in \lambda\mathscr{R}$$

by Lemma 6. Thus finally

$$\left(y_1, ..., y_J, \frac{\lambda}{\lambda'} a_{J+1}, ..., \frac{\lambda}{\lambda'} a_n\right) \in \mathscr{W}_J(\lambda),$$

where

$$y_j = \{b_j + x_j\}, \quad b_j = \left(\frac{\lambda}{\lambda'} - 1\right) a_j.$$

Lemma 10 now gives (15) at once if $x_1, ..., x_J$ run over all values such that $(x_1, ..., x_J, a_{J+1}, ..., a_n) \in \mathscr{W}_J(\lambda')$; and, as already remarked, this proves the lemma.

Proof of Theorem V (concluded). First,

$$V_n(\tfrac{1}{2}\lambda_1) = V_0(\tfrac{1}{2}\lambda_1) = (\tfrac{1}{2}\lambda_1)^n V$$

by (13). By Lemma 9 we have $V_n(\lambda) = V_1(\lambda)$ for $\tfrac{1}{2}\lambda_1 \leqslant \lambda \leqslant \tfrac{1}{2}\lambda_2$, and so

$$V_n(\tfrac{1}{2}\lambda_2) \geqslant (\lambda_2/\lambda_1)^{n-1} V_n(\tfrac{1}{2}\lambda_1)$$

by Lemma 11. Similarly,

$$V_n(\tfrac{1}{2}\lambda_3) \geqslant (\lambda_3/\lambda_2)^{n-2} V_n(\tfrac{1}{2}\lambda_2),$$
$$\vdots$$
$$V_n(\tfrac{1}{2}\lambda_n) \geqslant (\lambda_n/\lambda_{n-1})^1 V_n(\tfrac{1}{2}\lambda_{n-1}).$$

On multiplying these together,

$$V_n(\tfrac{1}{2}\lambda_n) \geqslant 2^{-n}\lambda_1 ... \lambda_n V.$$

This, with (12), gives $\lambda_1 ... \lambda_n V \leqslant 2^n$, as required.

THEOREM VI. (Mahler.) *There is a set of n integral points* $\mathbf{y}^{(r)}$ $(1 \leqslant r \leqslant n)$ *such that* $\det(y_{rj}) = \pm 1$ *and†* $V \Pi F(\mathbf{y}^{(r)}) \leqslant 2 . n!$.

† The constant $2.n!$ is not Mahler's and may clearly be further improved. What is important is that it depends only on n.

Proof. By Lemma 7 we may suppose that

$$\mathbf{x}^{(r)} = (x_{r1}, \ldots, x_{rr}, 0, \ldots, 0),$$

where $x_{rr} \neq 0$, since the $\mathbf{x}^{(r)}$ are linearly independent. We show that we may take

$$\mathbf{y}^{(r)} = (y_{r1}, \ldots, y_{r,\,r-1}, 1, 0, \ldots, 0)$$

for suitable integers $y_{r1}, \ldots, y_{r,\,r-1}$, such that $F(\mathbf{y}^{(r)}) \leqslant \mu_r$, where

$$\mu_1 = \lambda_1, \quad \mu_r = \tfrac{1}{2}r\lambda_r \quad (r \geqslant 2).$$

Theorem VI then follows from Theorem V.

Clearly $\mathbf{y}^{(1)} = x_{11}^{-1}\mathbf{x}^{(1)} = (1, 0, \ldots, 0)$ does what is required. Similarly if for any $r > 1$ we have $x_{rr} = \pm 1$ we may put

$$\mathbf{y}^{(r)} = x_{rr}^{-1}\mathbf{x}^{(r)},$$

which has integer co-ordinates and $F(\mathbf{y}^{(r)}) = \lambda_r \leqslant \mu_r$. We may thus suppose that

$$|x_{rr}| \geqslant 2.$$

There are certainly constants $\beta_1, \ldots, \beta_{r-1}$ such that

$$\mathbf{e}^{(r)} = (0, \ldots, 0, 1, 0, \ldots, 0) = \beta_1\mathbf{x}^{(1)} + \ldots + \beta_{r-1}\mathbf{x}^{(r-1)} + x_{rr}^{-1}\mathbf{x}^{(r)},$$

the 1 being in the rth place. Choose integers $b_1, \ldots, b_{r-1}$ so that $|\beta_j - b_j| \leqslant \tfrac{1}{2}$ and put

$$\begin{aligned}\mathbf{y}^{(r)} &= \mathbf{e}^{(r)} - b_1\mathbf{x}^{(1)} - \ldots - b_{r-1}\mathbf{x}^{(r-1)} \\ &= x_{rr}^{-1}\mathbf{x}^{(r)} + (\beta_1 - b_1)\mathbf{x}^{(1)} + \ldots + (\beta_{r-1} - b_{r-1})\mathbf{x}^{(r-1)}.\end{aligned}$$

Then, from the first expression, $\mathbf{y}^{(r)}$ has integer co-ordinates, and, from the second,

$$\begin{aligned}F(\mathbf{y}^{(r)}) \\ \leqslant |x_{rr}|^{-1}F(\mathbf{x}^{(r)}) + |\beta_1 - b_1|\,F(\mathbf{x}^{(1)}) + \ldots + |\beta_{r-1} - b_{r-1}|\,F(\mathbf{x}^{(r-1)}) \\ \leqslant \tfrac{1}{2}(\lambda_1 + \ldots + \lambda_r) \leqslant \tfrac{1}{2}r\lambda_r = \mu_r.\end{aligned}$$

NOTES

The proofs of Theorems V, VI are adapted from WEYL (1942). For an extension to arbitrary point sets see ROGERS (1949) and MAHLER (1049) or CHABAUTY (1949).

161

APPENDIX C

GAUSS'S LEMMA

LEMMA. (Gauss.) *Let $f=f(x_1, ..., x_m)$, $g=g(x_1, ..., x_m)$ be polynomials in any number of variables $x_1, ..., x_m$. Suppose that each of f, g has rational integer coefficients without common divisor. Then the coefficients of the product fg are integers without common divisor.*

Proof. We may write

$$f=\sum_I a_I I, \quad g=\sum_I b_I I, \tag{1}$$

where I runs through the monomials

$$I = x_1^{i_1} ... x_m^{i_m}.$$

Then

$$fg = \Sigma c_I I, \tag{2}$$

where

$$c_I = \sum_{JK=I} a_J b_K. \tag{3}$$

We shall say that $I = x_1^{i_1} ... x_m^{i_m}$ is lower than $J = x_1^{j_1} ... x_m^{j_m}$ if the first in order of the differences $j_1 - i_1, ..., j_m - i_m$ which does not vanish is positive. If $IJ = I_0 J_0$ then clearly either (i) $I = I_0$, $J = J_0$ or (ii) I is lower than I_0 or (iii) J is lower than J_0.

Let p be any prime. By hypothesis p does not divide all the a_I. Let I_0 be the lowest monomial such that $p \nmid a_{I_0}$. Similarly there is a lowest monomial J_0 such that $p \nmid b_{J_0}$. Then

$$c_{I_0 J_0} = \Sigma a_I b_J \quad (IJ = I_0 J_0). \tag{4}$$

If I is lower than I_0 then $p \mid a_I$ and if J is lower than J_0 then $p \mid b_J$, by hypothesis. Hence p divides all the summands in (4) except $a_{I_0} b_{J_0}$, which it does not divide. Thus

$$p \nmid c_{I_0 J_0}, \quad p \nmid \text{g.c.d. } (c_I).$$

Since p is any prime, this proves the lemma.

COROLLARY 1. *Suppose that the coefficients of f, g are no longer restricted to be without common divisor. Then*

$$\text{g.c.d. } (c_I) = \text{g.c.d. } (a_I) \cdot \text{g.c.d. } (b_I).$$

Proof. Consider $(\text{g.c.d. } (a_I))^{-1} f$, $(\text{g.c.d. } (b_I))^{-1} g$ instead of f, g.

II   CDA

COROLLARY 2. *Let* $f(x_1, \ldots, x_m)$ *have integer coefficients without common divisor and let* $g(x_1, \ldots, x_m)$ *have rational coefficients. If fg has integer coefficients then the coefficients of g are in fact integers.*

Proof. Let t be an integer such that the coefficients of tg are integers. Then $t \cdot fg = f \cdot tg$ has integer coefficients all divisible by t, by hypothesis; so

$$t \mid \text{g.c.d.} \, (a_I) \, \text{g.c.d.} \, (tb_I)$$

by the previous corollary. Since g.c.d. $(a_I) = 1$, by hypothesis, the b_I must be integers, as asserted.

COROLLARY 3. *Suppose that* f, g *have rational coefficients and that fg has integer coefficients. Then there is a rational number* k *such that* kf, $k^{-1}g$ *both have integer coefficients.*

Proof. There is certainly a rational number k such that the coefficients of kf are integers without common divisor. Since $fg = (kf) (k^{-1}g)$, the previous corollary applies to kf, $k^{-1}g$.

BIBLIOGRAPHY

BARNES, E. S. (1956). On linear inhomogeneous Diophantine approximation. *J. Lond. Math. Soc.* **31**, 73–9.

BARNES, E. S. & SWINNERTON-DYER, H. P. F. (1952). The inhomogeneous minima of binary quadratic forms I, II. *Acta Math., Stockh.* **87**, 259–323; **88**, 279–316.

BARNES, E. S. & SWINNERTON-DYER, H. P. F. (1955). The inhomogeneous minima of binary quadratic forms III. *Acta Math., Stockh.* **92**, 199–234.

BIRCH, B. J. (1956). A transference theorem of the geometry of numbers. *J. Lond. Math. Soc.* **31**, 248–51.

BIRCH, B. J. (1957). Transference theorems of the geometry of numbers, II. To appear in *Proc. Camb. Phil. Soc.*

BLANEY, H. (1950). Some asymmetric inequalities. *Proc. Camb. Phil. Soc.* **46**, 359–76.

CASSELS, J. W. S. (1950a). Some metrical theorems in Diophantine approximation I. *Proc. Camb. Phil. Soc.* **46**, 209–18.

CASSELS, J. W. S. (1950b). Some metrical theorems in Diophantine approximation III. *Proc. Camb. Phil. Soc.* **46**, 219–25.

CASSELS, J. W. S. (1950c). Some metrical theorems in Diophantine approximation IV. *Proc. K. Ned. Akad. Wet. Amst.* **53**, 176–87 (= *Indag. Math.* **12**, 14–25).

CASSELS, J. W. S. (1951). Some metrical theorems in Diophantine approximation V. On a conjecture of Mahler. *Proc. Camb. Phil. Soc.* **47**, 18–21.

CASSELS, J. W. S. (1952a). The product of n inhomogeneous linear forms in n variables. *J. Lond. Math. Soc.* **27**, 485–92.

CASSELS, J. W. S. (1952b). The inhomogeneous minimum of binary quadratic, ternary cubic and quaternary quartic forms. *Proc. Camb. Phil. Soc.* **48**, 72–86, 519–20.

CASSELS, J. W. S. (1953). A new inequality with application to the theory of Diophantine approximation. *Math. Ann.* **126**, 108–18.

CASSELS, J. W. S. (1954a). Über $\lim\limits_{x \to +\infty} x \mid \theta x + \alpha - y \mid$. *Math. Ann.* **127**, 288–304.

CASSELS, J. W. S. (1954b). On the product of two inhomogeneous forms. *J. reine angew. Math.* **193**, 65–83.

CASSELS, J. W. S. (1955). Simultaneous Diophantine approximation II. *Proc. Lond. Math. Soc.* (3), **5**, 435–48.

CASSELS, J. W. S. & SWINNERTON-DYER, H. P. F. (1955). On the product of three homogeneous linear forms and indefinite ternary quadratic forms. *Phil. Trans.* A, **248**, 73–96.

CHABAUTY, C. (1949). Sur les minima arithmétiques des formes. *Ann. Sci. Éc. Norm. Sup., Paris,* (3), **66**, 367–94.

164   DIOPHANTINE APPROXIMATION

CHABAUTY, C. & LUTZ, E. (1950). Sur les approximations linéaires réelles. *C.R. Acad. Sci., Paris,* 231, 938–39.

COHN, H. (1955). Approach to Markoff's minimal forms through modular functions. *Ann. Math., Princeton,* (2), 61, 1–12.

VAN DER CORPUT, J. G. (1931). Diophantische Ungleichungen. I, Zur Gleichverteilung modulo Eins. *Acta Math., Stockh.* 56, 373–456. II, Rythmische Systeme A und B. *Acta Math., Stockh.* 59, 209–328 (1932). (Promised parts C and D have not yet appeared.)

DAVENPORT, H. (1954). Simultaneous Diophantine approximation. *Proceedings International Conference of Mathematicians, Amsterdam,* 3, 9–12.

DAVENPORT, H. (1955). On a theorem of Furtwängler. *J. Lond. Math. Soc.* 30, 186–95.

DAVENPORT, H. & ROTH, K. F. (1955). Rational approximations to algebraic numbers. *Mathematika,* 2, 160–7.

DAVIS, C. S. (1950). The minimum of an indefinite binary quadratic form. *Quart. J. (Oxford),* (2), 1, 241–2.

DICKSON, L. E. (1930). *Studies in the theory of numbers* (especially Chapter VII): Chicago Univ. Press.

DUFRESNOY, J. & PISOT, C. (1953). Sur un ensemble fermé d'entiers algébriques. *Ann. Sci. Éc. Norm. Sup., Paris,* (3), 70, 105–34.

ERDŐS, P. & TURÁN, P. (1948). On a problem in the theory of uniform distribution (especially Theorem III). *Proc. K. Ned. Akad. Wet. Amst.* 51, 1146–54, 1262–9 (= *Indag. Math.* 10, 370–82, 406–13).

FROBENIUS, G. (1913). Über die Markoffschen Zahlen. *Preuss. Akad. Wiss. Sitzungsberichte,* 458–87.

GEL'FOND, A. O. (1941). (Гельфонд, А. О.) О дробных долях линейных комбинаций полиномов и показательных функций. *Мат. Сборник (Нов. Сер.)* 9, 721–6.

GEL'FOND, A. O. (1952). (Гельфонд, А. О.) *Трансцендентные и алгебраические числа.* Москва.

HALL, M. (1947). On the sum and product of continued fractions. *Ann. Math., Princeton,* (2), 48, 966–93.

HARDY, G. H., LITTLEWOOD, J. E. & PÓLYA, G. (1934). *Inequalities.* Cambridge: 2nd edition, 1952.

HARDY, G. H. & WRIGHT, E. M. (1938). *The Theory of Numbers.* Oxford: 3rd edition, 1954.

HLAWKA, E. (1952). Zur Theorie des Figurengitters. *Math. Ann.* 125, 183–207.

HLAWKA, E. (1954a). Zur Theorie der Überdeckung durch konvexe Körper. *Monatshefte Math. Phys.* 58, 287–91.

HLAWKA, E. (1954b). Inhomogene Minima von Sternkörpern. *Monatshefte Math. Phys.* 58, 292–305.

JARNÍK, V. (1946). Sur les approximations Diophantiques linéaires non homogènes. *Bull. Intern. de l'Acad. Tchèque des Sciences,* 16.

JARNÍK, V. (1954). (Ярник, В.) К теории однородных линейных
Диофантовых приближений. *Чехословацкий Мат. Журн.* 4, (79),
330–53 (with French summary).

KANAGASABAPATHY, P. (1952). Note on Diophantine approximation.
Proc. Camb. Phil. Soc. 48, 365–6.

KELLY, J. B. (1950). A closed set of algebraic integers. *Amer. J. Math.*
72, 565–72.

KHINTCHINE,* A. YA. (1923). Ein Satz über Kettenbrüche mit arith-
metischen Anwendungen. *Math. Z.* 18, 289–306.

KHINTCHINE, A. YA. (1926). Über eine Klasse linearer Diophantischer
Approximationen. *Rendiconti Circ. Mat. Palermo,* 50, 170–95.

KHINTCHINE, A. YA. (1935). (Хинчин, А. Я.) *Цепные дроби.* Москва—
Ленинград. 2nd edition, 1949.

KHINTCHINE, A. YA. (1948a). (Хинчин, А. Я.) Количественная
концепция аппроксимацонной теории Кронекера. *ИзвестияАкад.
Наук С.С.С.Р. (Сер. Мат.),* 12, 113–22.

KHINTCHINE, A. YA. (1948b). (Хинчин, А. Я.) Регулярные системы
линейных уравненний и общая задача Чевышева. *Известия
Акад. Наук С.С.С.Р. (Сер. Мат.),* 12, 249–58.

KNESER, M. (1955). Ein Satz über abelsche Gruppen mit Anwendungen
auf die Geometrie der Zahlen. *Math. Z.* 61, 429–34.

KOKSMA, J. F. (1936). *Diophantische Approximationen.* Ergebnisse d.
Math. u. ihrer Grenzgebiete 4, 4. Berlin und Leipzig.

KOKSMA, J. F. (1937). Über einen Dirichlet-Minkowskischen Approxi-
mationssatz. *Mathematica B, Zutphen,* 6, 113–31, 171–81.

LANDAU, E. (1927). *Vorlesungen über Zahlentheorie* (3 Bände). Leipzig.

LEVEQUE, W. J. (1953). Note on S-numbers. *Proc. Amer. Math. Soc.*
4, 189–90.

LUTZ, É. (1951). Sur les approximations diophantiennes linéaires
P-adiques. Thèse, Strasbourg (=*Actualités Sci. Ind.* 1224, 1955).

MAHLER, K. (1939a). Ein Übertragungsprinzip für lineare Ungleichungen. *Čas. Pěst. Mat.* 68, 85–92.

MAHLER, K. (1939b). Ein Übertragungsprinzip für konvexe Körper.
Čas. Pěst. Mat. 68, 93–102.

MAHLER, K. (1946). On lattice points in *n*-dimensional star bodies.
I. Existence theorems. *Proc. Roy. Soc. A,* 187, 151–87.

MAHLER, K. (1949). On the minimum determinant of a special point set.
Proc. K. Ned. Akad. Wet. Amst. 52, 633–42 (=*Indag. Math.* 11,
195–204).

MAHLER, K. (1953a). On the approximation of logarithms of algebraic
numbers. *Phil. Trans. A,* 245, 371–98.

MAHLER, K. (1953b). On the approximation of π. *Proc. K. Ned. Akad.
Wet. Amst. A,* 56, (=*Indag. Math.* 15), 29–42.

MAHLER, K. (1955). On compound convex bodies I, II. *Proc. Lond.
Math. Soc.* (3), 5, 358–84.

---

* Transliterated Hinčin by *Math. Rev.*

MARKOFF, A. (1879). Sur les formes quadratiques binaires indéfinies. *Math. Ann.* **15**, 381–409.

MORDELL, L. J. (1951). On the product of two non-homogeneous linear forms, IV. *J. Lond. Math. Soc.* **26**, 93–5.

PERRON, O. (1913). *Die Lehre von den Kettenbrüchen.* Leipzig und Berlin; 3rd edition, 1954, Stuttgart.

PERRON, O. (1921). *Irrationalzahlen.* Berlin und Leipzig.

PISOT, C. (1946). Répartition (mod 1) des puissances successives des nombres réels. *Comm. Math. Helv.* **19**, 153–60.

POITOU, G. (1953). Sur l'approximation des nombres complexes par les nombres des corps imaginaires quadratiques, etc. *Ann. Sci. Éc. Norm. Sup., Paris*, (3), **70**, 199–265.

PONTRJAGIN, L. S. (1938). (Понтрягин, Л. С.). *Непрерывные группы.* Москва. 2nd Edition, 1954. There is an American translation: *Topological groups* (Princeton, 1939).

REMAK, R. (1924). Über indefinite binäre quadratische Minimalformen. *Math. Ann.* **92**, 155–82.

REMAK, R. (1925). Über die geometrische Darstellung der indefiniten binären quadratischen Minimalformen. *Jber. Dtsch. MatVer.* **33**, 228–45.

ROGERS, C. A. (1949). The product of the minima and the determinant of a set. *Proc. K. Ned. Akad. Wet. Amst.* **52**, 256–63 (= *Indag. Math.* **11**, 71–8).

ROGERS, C. A. (1954). The product of *n* non-homogeneous linear forms. *Proc. Lond. Math. Soc.* (3), **4**, 50–83.

ROTH, K. F. (1954). On irregularities of distribution. *Mathematika*, **1**, 73–9.

ROTH, K. F. (1955). Rational approximations to algebraic numbers. *Mathematika*, **2**, 1–20 (with corrigendum p. 168).

SAMET, P. A. (1953). Algebraic integers with two conjugates outside the unit circle I, II. *Proc. Camb. Phil. Soc.* **49**, 421–36 and **50**, 346 (1954).

SAWYER, D. B. (1950). A note on the product of two non-homogeneous linear forms. *J. Lond. Math. Soc.* **25**, 239–40.

SCHNEIDER, T. (1956). *Einführung in die transzendenten Zahlen.* Berlin, Göttingen und Heidelberg.

SEGRE, B. (1945). Lattice points in infinite domains and asymmetric diophantine approximations. *Duke Math. J.* **12**, 337–65.

SIEGEL, C. L. (1949). *Transcendental numbers* (*Annals of Mathematics Studies* **16**). Princeton Univ. Press.

TORNHEIM, L. (1955). Asymmetric minima of quadratic forms and asymmetric diophantine approximation. *Duke Math. J.* **22**, 287–94.

TURÁN, P. (1953). *Eine neue Methode in der Analysis und deren Anwendungen.* Budapest. There is a rather expanded version in Hungarian published simultaneously. A new improved edition is appearing in Chinese.

VINOGRADOFF, I. M. (1947). (Виноградов, И. М.) *Метод Тригоно-метрических сумм в теории чисел.* Труды МАТ. ИНСТ. ИМ. В. А. СТЕКОВА XXIII, reprinted in his *Избранные труды* (Москва, 1952). There is an expanded English translation: *The method of trigono-metrical sums in the theory of numbers* (London and New York, ?1954), with explanatory notes, references to later literature and not so many misprints.

WEYL, H. (1942). On geometry of numbers. *Proc. Lond. Math. Soc.* (2), 47, 268–89.

SUPPLEMENTARY BIBLIOGRAPHY

BAKER, A. (1964*). Rational approximations to certain algebraic numbers. *Proc. London Math. Soc.* (3) 14, 385–98.

BIRCH, B. J. (1957*). *The geometry of numbers.* Cambridge Ph.D. thesis. (Especially Chapter III: Inhomogeneous minima of complex quadratic forms, still, alas! unpublished).

CASSELS, J. W. S. (1959*). *An introduction to the geometry of numbers.* Springer, Berlin.

CIGLER, J. & HELMBERG, G. (1961*). Neuere Entwickelungen der Theorie der Gleichverteilung. *Jahresberich der DMV*, 64, 1–50.

DAVENPORT, H. (1962*). A note on diophantine approximation, I, II. *Studies in mathematical analysis and related topics*, 77–81 (Stanford University Press) and *Mathematika* 11 (1964), 50–8.

DAVENPORT, H., ERDÖS, P. & LEVEQUE, W. J. (1963*). On Weyl's criteria for uniform distribution. *Mich. Math. J.* 10, 311–4.

DAVENPORT, H. & LEVEQUE, W. J. (1963*). Uniform distribution relative to a fixed sequence. *Mich. Math. J.* 10, 315–9.

DESCOMBES, R. (1956*). Sur la répartition des sommets d'une ligne polygonale régulière non fermée. *Ann. Ecole Norm. Sup.* (3) 73, 285–355.

EGGAN, L. C. (1961*). On diophantine approximations. *Trans. Amer. Math. Soc.* 99, 102–17.

EGGAN, L. C. & NIVEN, I. (1961*). A remark on one-sided approxima-tion. *Proc. Amer. Math. Soc.* 12, 538–40.

EGGLESTON, H. G. (1951*). Sets of fractional dimension which occur in some problems of number theory. *Proc. London Math. Soc.* (2) 54, 42–93.

ENNOLA, V. (1958). On the first inhomogeneous minimum of indefinite quadratic forms and Euclid's algorithm in real quadratic fields. *Ann. Univ. Turkuensis, Ser.* AI, 28, 9–58.

ERDÖS, P. (1959*). Some results on diophantine approximation. *Acta Arith.* 5, 359–69.

GALLAGHER, P. (1961*). Approximation by reduced fractions. *J. Math. Soc. Japan*, **13**, 342–5.

GALLAGHER, P. (1962*). Metric simultaneous diophantine approximations. *J. London Math. Soc.* **37**, 387–90.

GELFOND, A. O. (1960*). *Transcendental and algebraic numbers*. New York, Dover.

NIVEN, I. (1963*). *Diophantine approximations*. New York, Interscience.

KAC, M. (1959*). *Statistical independence in probability, analysis and number-theory*. Carus Monograph No. 14. New York, Wiley.

KHINCHIN, A. YA. (1964*). *Continued fractions*. Chicago Univ. Press.

LANG, S. (1962*). *Diophantine geometry*. New York, Interscience.

LEVEQUE, W. J. (1956*). *Topics in number-theory*. 2 Vols. Reading, Mass., Addison-Wesley.

LEVEQUE, W. J. (1958*). On the frequency of small fractional parts in certain real sequences. *Trans. Amer. Math. Soc.* **87**, 237–360 and **94** (1959), 130–49.

MAHLER, K. (1963*). On the approximation of algebraic numbers by algebraic integers. *J. Austral. Math. Soc.* **3**, 408–34.

MAHLER, K. (1961*). *Lectures on diophantine approximation*. Univ. of Notre Dame Press.

PECK, L. G. (1961*). Simultaneous rational approximations to algebraic numbers. *Bull. Amer. Math. Soc.* **67**, 197–201.

PERRON, O. (1946*). Ein Analogon zu einem Satz von Minkowski. *Sitz. d. Bayr. Akad. d. Wiss.* **1946**, 159–65.

PITMAN, J. (1960*). Davenport's constant for indefinite quadratic forms. *Acta Arith.* **6**, 37–46.

PISOT, C. (1962*). Eine merkwürdige Klasse ganzer algebraischer Zahlen. *J. reine angew. Math.* **209**, 82–3.

POSTNIKOV, A. G. (1960*). (Постников А.Г.) Арифметическое Моделирование случайных лроцессов. *Трчды Мат. Инст. им. Стеклова,* **57**, 1–84.

ROGERS, C. A. (1964*). Some sets of continued fractions. *Proc. London Math. Soc.* (3) **14**, 29–44.

RYLL-NARDZEWSKI, C. (1951*). On the ergodic theorems, II. Ergodic theory of continued fractions. *Studia Math.* **12**, 74–9.

SALEM, R. (1963*). *Algebraic numbers and Fourier analysis*. Boston, Mass., Heath.

SCHMIDT, W. (1960*). A metrical theorem in diophantine approximation. *Canad. J. Math.* **12**, 619–631.

SZÜSS, P. (1958*). Über die metrische Theorie der diophantischen Approximationen, I, II. *Acta Math. Hung.* **9**, 179–93 and *Acta Arith.* **8** (1963), 225–41.

INDEX OF DEFINITIONS

These words and phrases are set in SMALL CAPITALS where they are defined.